Synthesis Lectures on Engineering, Science, and Technology

The focus of this series is general topics, and applications about, and for, engineers and scientists on a wide array of applications, methods and advances. Most titles cover subjects such as professional development, education, and study skills, as well as basic introductory undergraduate material and other topics appropriate for a broader and less technical audience.

Kevin M. P. Ryan

Polynomial Symmetry in Butterworth Filter Synthesis

Exact Design of Doubly-Terminated Passive Ladder Networks

Springer

Kevin M. P. Ryan
School of Engineering and Applied Science
Yale University
New Haven, CT, USA

ISSN 2690-0300 ISSN 2690-0327 (electronic)
Synthesis Lectures on Engineering, Science, and Technology
ISBN 978-3-032-33467-1 ISBN 978-3-032-33468-8 (eBook)
https://doi.org/10.1007/978-3-032-33468-8

This Springer imprint is published by the registered company Springer Nature Switzerland AG
The registered company address is: Gewerbestrasse 11, 6330 Cham, Switzerland

To Frogard,
with love and gratitude.

Preface

This book grew out of a question that stayed with me for many years: how do passive ladder filters actually produce the responses we assign to them?

Early in my undergraduate education, I paid to attend a course in analog filter design. I expected to come away with a deeper understanding of filters, especially how their circuit structures produced the frequency responses used so widely in practice. The material I encountered was useful, but it was largely procedural. The design process moved through normalized prototypes, polynomial forms, tables, transformations, and scaling rules. The results were correct, and the methods were certainly efficient, but the underlying circuit relationships remained partly hidden. I learned how to obtain element values, but the connection between those values and the passive network itself remained less clear than I had hoped.

That dissatisfaction stayed with me through a career that included practical filter design. Years of practical design work and teaching-related engineering work gave me the opportunity to return to this problem with more experience, better tools, and a clearer sense of what had been missing. Over time, the issue became less a criticism of conventional methods than a persistent technical and pedagogical question. The standard approaches to Butterworth filter design are powerful and well established. They are also compact, which is one reason they have endured. However, compactness can come at the cost of visibility. When the design is presented primarily as a sequence of normalized tables and polynomial manipulations, the relationship between the ladder network, the source and load terminations, the cutoff condition, and the Butterworth coefficient structure can be difficult to see.

This book is my attempt to make those relationships explicit.

The treatment developed here begins with passive ladder networks and their terminations rather than with a completed prototype table. The passband level is set by the source-load voltage divider, and the cutoff level is defined relative to that passband level. At cutoff, the frequency-independent and frequency-dependent contributions to the denominator become equal. This equality provides the starting point for connecting the circuit equations to the normalized Butterworth polynomial. Once that relationship is made explicit, the synthesis problem becomes a matter of matching the appropriate Butterworth coefficient structure to the transfer-function terms produced by the chosen ladder network.

The emphasis throughout the book is on low-pass and high-pass passive Butterworth ladders, with both shunt-first and series-first topologies. Time constants, source and load resistance, coefficient matching, product-factor relationships, and topology-dependent element placement are used to show how the component values arise from the circuit equations. The objective is not to replace the classical normalized-prototype method, but to open it up and show more of the structure underneath it.

The intended readers are students, instructors, and practicing engineers who want more than a table-driven design procedure. Readers who already know conventional filter design methods may find this book useful as a complementary derivation. Readers encountering passive ladder synthesis for the first time may find that the step-by-step development gives the circuit equations a clearer role in the design process.

The book proceeds from lower-order filters to higher-order examples, allowing the algebraic structure to develop gradually. First- and second-order cases establish the basic denominator relationships. Third- and fourth-order cases show how the method extends while preserving direct connections between topology and coefficient matching. Fifth- and sixth-order cases illustrate both the power and the increasing complexity of the approach. The later chapters and appendices are intended to preserve enough detail that readers can reproduce the calculations rather than simply accept the final element values.

I wrote this book for the kind of reader I once was: someone willing to work through the equations, but wanting those equations to reveal the circuit relationships rather than conceal them. My hope is that the development presented here helps make passive Butterworth ladder design feel less like a sequence of recipes and more like a connected synthesis process.

New Haven, CT, USA Kevin M. P. Ryan
June 2026

Acknowledgements The author used ChatGPT as a non-authorial editorial and technical-assistance tool during final manuscript preparation, including copyediting support, consistency review, LaTeX troubleshooting, compile-warning cleanup, revision tracking, and preliminary index-term planning. The author reviewed, verified, and approved the final manuscript and remains solely responsible for its content.

Competing Interests The author has no competing interests to declare that are relevant to the content of this manuscript.

Contents

Part I
Fundamentals

Abstract This chapter introduces the motivation, scope, and practical setting of the monograph. It places Butterworth filter design in historical context, identifies limitations of approximation-based workflows, and explains why doubly-terminated passive ladder networks provide a natural framework for accurate synthesis. The chapter also motivates the explicit treatment of source and load resistances as design parameters rather than incidental boundary conditions and prepares the structural developments of the chapters that follow.

1 Introduction

Exactness tends to be exacting.

The Butterworth filter was introduced by Stephen Butterworth in his 1930 paper, "On the Theory of Filter Amplifiers" [1]. Nearly a century later, Butterworth responses remain among the most widely used in analog signal processing because of their maximally flat passband characteristic and their practical usefulness in passive and active filter design. Despite their ubiquity, however, many standard design workflows still depend on normalized prototypes, tabulated values, scaling rules, or software tools that obscure the mathematical structure of the underlying synthesis problem.

This monograph revisits Butterworth filter design from an analytical perspective. Its principal aim is to show that the denominator of a doubly-terminated passive low-pass ladder network can be matched, degree by degree, to the coefficient structure of the corresponding normalized Butterworth polynomial. In that sense, the work is both retrospective and constructive: it returns to the original problem of Butterworth synthesis while seeking a more transparent and exact design framework for contemporary use.

Traditional hand methods were developed in an era of tables, nomographs, and slide rules. Modern practice has largely replaced those tools with numerical routines, calculators, and web-based utilities. These tools are convenient, but they often conceal the dependence of the final design on source resistance, load resistance, cutoff frequency, and ladder topology. That concealment becomes especially problematic when the source and load resistances are unequal, since many common workflows either assume equal terminations or handle unequal terminations only approximately.

K. M. P. Ryan, *Polynomial Symmetry in Butterworth Filter Synthesis*,
Synthesis Lectures on Engineering, Science, and Technology,
https://doi.org/10.1007/978-3-032-33468-8_1

A familiar example appears in Appendix E of *The Art of Electronics* [2], where a simplified procedure based on Zverev's work [3] is presented. That approach is useful and practical, but it remains approximate and still relies on tabulated information. In particular, when the source and load resistances are close in value but not equal, approximation-based methods may introduce noticeable error.

The approach developed in this monograph follows a different principle: it is preferable to begin with an exact mathematical description and then approximate deliberately, if necessary, than to begin with an approximation and attempt to recover precision afterward. The purpose of this chapter is to establish the problem setting in which that principle matters and to explain why doubly-terminated passive low-pass ladder networks provide the appropriate framework for the synthesis method that follows.

1.1 Brief History

Butterworth's 1930 paper introduced what is now recognized as the maximally flat passband filter. Since then, Butterworth responses have become standard tools in analog circuit design, and a substantial practical tradition has grown around their use. Much of that tradition, however, has emphasized implementation procedure over analytical transparency. Designers have often relied on prototype tables, interpolation formulas, scaling rules, or software utilities that produce usable component values without making the governing relationships especially clear.

That practical tradition has obvious value: it has enabled efficient filter design for generations of engineers. Nevertheless, when unequal source and load resistances must be treated explicitly, or when exact element values are desired before approximation to preferred component series, many familiar workflows become awkward. This monograph addresses that difficulty by presenting a synthesis procedure that yields exact capacitor and inductor values for a specified cutoff frequency with either equal or unequal terminations.

In practice, exact values will rarely coincide with standard component series. Even so, an exact result is the better starting point. Once the exact solution is known, approximation to available components becomes a controlled engineering choice rather than an inherited limitation of the design method itself.

1.2 Ladder Topology

The ladder network is one of the most important topologies in passive analog filter design. It may be realized in asymmetrical or symmetrical forms and is closely associated with the work of Wilhelm Cauer, although related ladder constructions also appear in earlier work by George Campbell. Because ladder networks provide direct passive realizations with clear physical interpretations, they form the natural setting for the present development.

The methods presented in this monograph are introduced in the context of ladder filters because that context is both practical and analytically convenient. The underlying transformations, however, are not conceptually restricted to ladders. More generally, they apply to networks whose behavior can be represented by an appropriate transfer function and whose element relationships can be recovered from the associated polynomial structure.

1.2.1 Why Doubly-Terminated

A passive filter does not operate in isolation. At its input, it is driven by a source whose nonzero output resistance appears as the filter's source resistance, R_S. At its output, it feeds a receiving circuit whose finite input resistance appears as the filter's load resistance, R_L. These resistances are not merely incidental parasitics; they participate directly in determining the realized cutoff frequency and overall response of the network.

For that reason, an accurate passive filter design should treat both terminations explicitly. This is the motivation for the doubly-terminated ladder topology used throughout this monograph. By retaining both R_S and R_L in the formulation, the design remains faithful to the electrical environment in which the filter will actually operate.

Many classical workflows assume equal source and load resistances from the outset. In other cases, one termination is neglected entirely, or the filter is treated as though it were driven from an ideal source and connected to an infinite load impedance. Such assumptions may be acceptable in special cases, but they are not generally appropriate when the goal is accurate realized performance or when the surrounding circuitry imposes unequal terminations.

The synthesis method developed here therefore makes no requirement that $R_S = R_L$. Instead, both resistances are carried explicitly through the design process. This allows unequal terminations to be handled naturally rather than approximately or as an afterthought.

1.2.2 Doubly-Terminated Example

Figures 1.1 and 1.2 illustrate the practical importance of explicit termination. In Fig. 1.1, the effective source resistance R_{SOURCE} and load resistance R_{LOAD} are determined largely by the surrounding active circuitry, in this case operational amplifiers. Those effective resistances depend on semiconductor device behavior and can vary with process, bias conditions, temperature, and supply voltage. As a result, they are often difficult to characterize accurately.

In Fig. 1.1, R_{SOURCE} is dominated primarily by the output resistance of U_1, together with the effect of the surrounding feedback network. Likewise, R_{LOAD} depends primarily on the input resistance presented by U_2, which may be large but still finite and variable. When

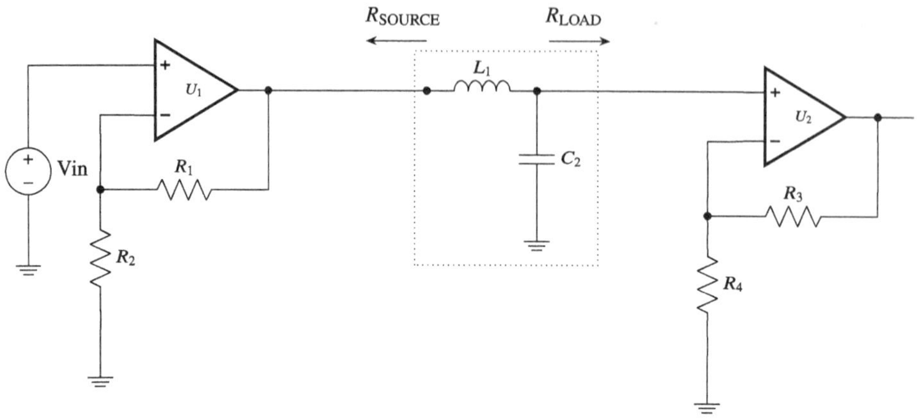

Fig. 1.1 Poorly characterized effective source and load resistances in a practical interface

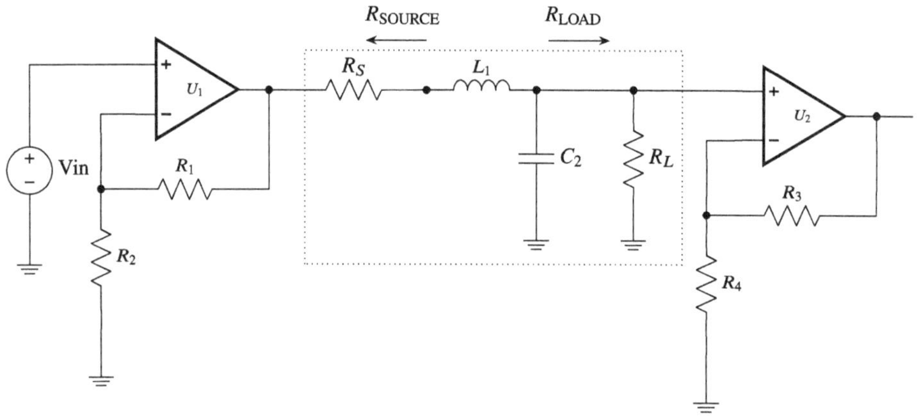

Fig. 1.2 Use of explicit termination resistors to define source and load conditions

these effective resistances are uncertain, the filter is being designed against poorly defined boundary conditions.

Figure 1.2 shows a more controlled approach. By inserting a series resistor R_S whose value is large compared with the output resistance of U_1, the effective source resistance seen by the filter becomes dominated by a known, intentional component. Similarly, by adding a shunt resistor R_L whose value is small compared with the input resistance of U_2, the effective load resistance becomes dominated by another known, intentional component. In both cases, the uncertain semiconductor contribution is reduced to a smaller error term.

This has two important advantages. First, the filter can be synthesized using well-characterized source and load resistances rather than uncertain effective values. Second, the added resistors can be chosen with known tolerances and temperature coefficients, making the realized filter behavior easier to predict across operating conditions.

Accordingly, once the source circuitry and receiving circuitry have been examined, it may be advantageous to introduce explicit termination resistors R_S and R_L as part of the filter design itself. These resistors need not be equal. Their role is not merely one of convenience, but of control: they define the electrical boundary conditions under which the passive ladder is intended to realize the target Butterworth response.

1.3 Scope of the Monograph

The chapters that follow are organized around a single analytical idea: the denominator of a doubly-terminated low-pass passive ladder transfer function can be brought into correspondence with the coefficient structure of the appropriate normalized Butterworth polynomial. The task of the early chapters is to identify the structural ingredients that make that correspondence possible. The task of the later chapters is to apply the resulting transformation rule to filters of increasing order.

The monograph therefore proceeds from interpretation to synthesis. It begins by identifying the physical and mathematical setting of the problem. It then turns to the symmetry of Butterworth polynomials, the interpretation of cutoff relative to the realized passband level, the transformation rule that maps realized low-pass ladder denominators to Butterworth coefficients, and the resulting solution structure. Only after that foundation is established does the discussion turn to order-specific network synthesis.

1.4 Transition to Chap. 2

The purpose of this chapter has been to establish why doubly-terminated passive ladder filters provide the right setting for exact Butterworth synthesis. The next step is to identify the structural property that makes such synthesis possible. That property is symmetry: first in the coefficient structure of the normalized Butterworth polynomial, and then in the degree-ordered low-pass denominator of the realized ladder network.

Accordingly, Chap. 2 examines the symmetry of Butterworth polynomials and the way that symmetry is projected into low-pass passive ladder transfer functions. Once that structure is understood, the cutoff condition and the transformation rule can be developed on a firmer analytical basis.

References

1. Butterworth, S. (1930). On the theory of filter amplifiers. *Experimental Wireless and the Wireless Engineer, 7*, 536–541.
2. Horowitz, P., & Hill, W. (2015). *Art of electronics* (3rd ed.). Cambridge University Press. ISBN 978-0-521-80926-9 Hardback.
3. Zverev, A. I. (1967). *Handbook of filter synthesis*. Wiley.

2 Symmetry

DC is to AC as the track is to a roller coaster.

Symmetry is one of the organizing principles of this monograph. In the present context, the term does not primarily refer to geometric mirror images, but to structural balance within polynomial forms and circuit realizations. In particular, the design method developed here depends on symmetry in two related senses: first, the symmetry of normalized Butterworth polynomial coefficients; and second, the symmetry of the relationships among the terms that appear in the transfer-function denominator of a doubly-terminated passive low-pass ladder filter.

This viewpoint is useful because it converts what is often treated as a table-lookup problem into a structural one. Once the relevant symmetry is recognized, the relationships among filter order, termination, and element values become easier to organize analytically. Symmetry alone does not solve the synthesis problem, but it reveals the pattern that makes exact synthesis possible.

2.1 Symmetry Definition

In mathematics, symmetry refers broadly to invariance or balanced correspondence under a transformation. In filter theory, the term may be used more narrowly to describe a structured relationship among coefficients or terms. For the purposes of this monograph, symmetry will mean that terms or coefficients may be paired about a center in a way that reveals a balanced underlying form.

K. M. P. Ryan, *Polynomial Symmetry in Butterworth Filter Synthesis*,
Synthesis Lectures on Engineering, Science, and Technology,
https://doi.org/10.1007/978-3-032-33468-8_2

This is especially important for Butterworth polynomials. Their coefficients are not arbitrary. They exhibit a predictable order-dependent structure, and that structure carries into the denominator forms of passive low-pass ladder transfer functions. The design method developed here exploits that correspondence.

Symmetry in this sense should not be confused with identity of every term. In an odd-order polynomial, the terms can often be arranged into complete pairs about a center. In an even-order polynomial, the same balancing tendency exists, but the middle structure is different. This distinction becomes important later when the implementation of even- and odd-order filters is considered separately.

2.1.1 Symmetry Example

A simple example may be seen in the normalized Butterworth polynomials of low order. For a third-order Butterworth polynomial,

$$s^3 + 2s^2 + 2s + 1,$$

the coefficients are symmetric in the sense that the outer coefficients match and the inner coefficients match:

$$1,\ 2,\ 2,\ 1.$$

For a second-order Butterworth polynomial,

$$s^2 + \sqrt{2}s + 1,$$

the outer coefficients remain matched, but there is now a single middle term:

$$1,\ \sqrt{2},\ 1.$$

These low-order examples illustrate the central point. Odd- and even-order Butterworth polynomials both exhibit structured balance, but that balance is expressed differently. The odd-order case admits complete pairing about the center, whereas the even-order case contains an unmatched middle contribution. That difference is not merely aesthetic; it reappears in the corresponding circuit forms.

2.2 Monotonicity

The design method presented in this monograph applies to doubly-terminated passive Butterworth filters in ladder form. Butterworth responses are especially suitable for this development because their magnitude behavior is monotonic in the passband-to-stopband transition: they do not exhibit ripple, and their defining characteristic is maximal flatness at zero frequency in the normalized low-pass case. The absence of ripple makes structural relationships easier to isolate, since the denominator is not shaped by oscillatory passband constraints of the kind found in Chebyshev or elliptic responses.

Monotonicity matters here because the transfer-function denominator is not merely a bookkeeping device. It governs how the realized magnitude changes with frequency, and its degree-by-degree structure can be related back to circuit elements and terminations. When that polynomial structure is also symmetric, the resulting synthesis problem becomes more tractable.

2.2.1 Log–Log Plots

Logarithmic plotting is often the clearest way to visualize monotonic filter behavior. On log-frequency axes, the transition from passband flatness to stopband roll-off becomes easier to interpret, and the influence of filter order on asymptotic slope is displayed compactly. Although the precise synthesis method developed here is algebraic rather than graphical, log-log or Bode-style views remain useful for understanding how polynomial order manifests as observable response shape.

Such plots also help distinguish local flatness from global roll-off. Near the origin, the Butterworth response is maximally flat. Farther from cutoff, the magnitude transitions into the familiar order-dependent attenuation behavior. The denominator polynomial encodes both of these features simultaneously.

2.2.2 Flatness, Linearity, and Slope

The Butterworth response is defined by passband flatness, not by piecewise linearity. Nevertheless, when plotted on logarithmic axes, the response develops visually interpretable slope behavior as frequency increases. This is useful because the order of the polynomial determines the eventual attenuation rate, while the coefficient structure governs how the response transitions into that rate.

For the present purpose, the important point is that monotonicity and symmetry coexist in the Butterworth case. The response is monotonic, while the denominator coefficients and associated transfer-function terms exhibit patterned balance. It is this combination that makes Butterworth filters especially amenable to the exact synthesis framework developed in the chapters that follow.

2.3 Symmetry in Monotonic Polynomials

This design method applies to doubly-terminated passive Butterworth filters in ladder form. The method is possible because of two related observations: the symmetry of normalized Butterworth polynomial coefficients, and the symmetry of the interrelationships among the degrees of the corresponding transfer-function polynomials.

These are not independent facts. The coefficient symmetry of the normalized Butterworth form is reflected in the structure of the denominator terms that arise from passive low-pass ladder realizations. In other words, the circuit polynomial inherits a recognizable projection of the symmetry already present in the target Butterworth polynomial. That inherited structure is what makes exact element extraction feasible.

2.3.1 Symmetry of Normalized Butterworth Polynomials

For odd-order normalized Butterworth polynomials, the coefficients exhibit complete pairing about the center. A third-order example is

$$s^3 + 2s^2 + 2s + 1.$$

Here the first and last coefficients are equal, and the two middle coefficients are equal. The coefficient sequence is therefore balanced about its center.

For even-order normalized Butterworth polynomials, the same outer symmetry remains, but a single middle coefficient stands alone. A second-order example is

$$s^2 + \sqrt{2}s + 1.$$

In this case the outer coefficients are equal, but the middle term is unmatched.

This odd-even distinction is basic to the design method. Odd-order forms are fully pairable about the center, whereas even-order forms are only partially pairable because of the unmatched middle degree. That asymmetry in pairing later determines the practical constraints on topology selection.

2.3.1.1 Butterworth Product Factors

Normalized Butterworth polynomial coefficients may be expressed in terms of Butterworth product factors. To generate those factors, this monograph primarily uses the product formula [1, 2]:

$$a_k = \prod_{\mu=1}^{k} \frac{\cos((\mu - 1)\gamma)}{\sin(\mu\gamma)}, \tag{2.1}$$

where

$$a_0 = 1 \quad \text{and} \quad \gamma = \frac{\pi}{2n}.$$

Equation (2.1) produces the Butterworth product factors from which the normalized Butterworth polynomial coefficients are constructed. In this form, the coefficients are generated from a rule rather than retrieved from a table, which is more consistent with the analytical goals of this monograph.

2.3.1.2 Butterworth Coefficients

Once the product factors are known, the normalized Butterworth coefficients follow in a structured way. What matters for the present discussion is not only their numerical value, but their ordered relationship. The coefficient sequence is neither arbitrary nor merely convenient; it embodies the balanced structure that the denominator of the realized passive low-pass network must reproduce.

Accordingly, the Butterworth coefficients serve two roles in the present method. First, they define the target response. Second, they reveal the symmetry that guides the arrangement and solution of the transfer-function terms.

2.3.2 Transfer Functions–Low-Pass Denominator

For doubly-terminated passive ladder filters, the low-pass denominator contains terms whose degree structure reflects the order of the network and whose coefficients depend on the ladder elements together with the source and load resistances. When arranged by degree, these terms reveal a symmetry that parallels the normalized Butterworth coefficients.

This is the essential connection between the target polynomial and the realizable circuit: the denominator produced by the ladder network is not an unrelated algebraic object. Its term structure can be matched, degree by degree, to the Butterworth form, and the symmetry within that matching provides the basis for exact synthesis.

2.3.2.1 Projection of Butterworth Polynomial Symmetry

The symmetry of the interrelationships among the degrees of the transfer-function denominator is most easily seen through representative low-order examples.

For an odd-order transfer-function denominator, a representative form is

$$\omega^3 C_1 L_2 C_3 R_S + \omega^2 \left(C_1 L_2 \frac{R_S}{R_L} + L_2 C_3 \right) + \omega \left(C_1 R_S + \frac{L_2}{R_L} + C_3 R_S \right) + \left(\frac{R_S}{R_L} + 1 \right).$$

This form is balanced in the sense that the outer and inner groupings mirror one another structurally: the highest- and lowest-degree terms define the exterior of the polynomial, while the intermediate degrees form corresponding interior groupings.

For an even-order transfer-function denominator, a representative form is

$$\omega^2 C_1 L_2 \frac{R_S}{R_L} + \omega\left(C_1 R_S + \frac{L_2}{R_L}\right) + \left(\frac{R_S}{R_L} + 1\right).$$

Here the same general ordering is present, but the symmetry is less complete because the middle degree is unmatched. This parallels the behavior already observed in the normalized Butterworth coefficient sequences: odd-order forms admit full pairing, while even-order forms retain only partial pairing.

The important conclusion is that the symmetry of odd- and even-order transfer-function polynomials follows the symmetry of the corresponding Butterworth polynomial coefficients. This correspondence is what allows the normalized target response to be mapped into a passive ladder realization without surrendering analytical transparency.

2.3.2.2 Even Versus Odd

This design methodology requires that even- and odd-order filters be treated differently in both design and implementation.

Odd-order filters are structurally symmetric, so either shunt-first or series-first topologies may be used regardless of the relative values of R_S and R_L. Their symmetry provides sufficient flexibility that the choice of starting element does not impose the same constraint seen in the even-order case.

Even-order filters are different. Because of their asymmetric pairing structure, topology selection must respect the relative magnitudes of the terminations. When $R_S > R_L$, a shunt-first implementation should be used, as in Fig. 2.1a. When $R_S < R_L$, a series-first implementation should be used, as in Fig. 2.1b. In the special case $R_S = R_L$, either shunt-first or series-first may be used.

A practical way to remember this rule is as follows: for even-order filters, place the capacitor in parallel with the larger resistance and the inductor in series with the smaller resistance. This mnemonic is not the proof of the rule, but it is a useful implementation guide.

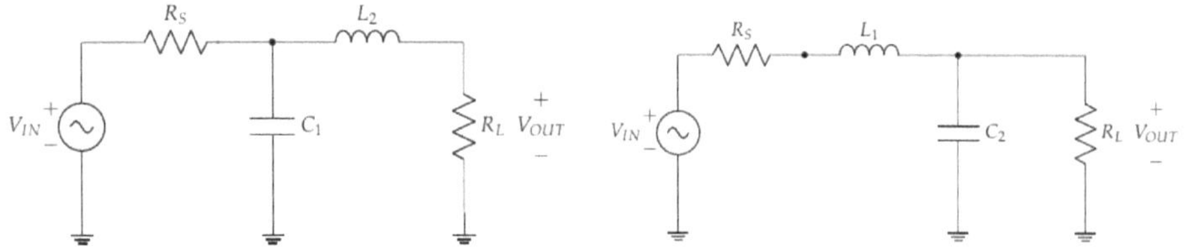

(a) Shunt-first for $R_S > R_L$ (b) Series-first for $R_S < R_L$

Fig. 2.1 Examples of even-order filter implementation

The distinction between even- and odd-order forms is therefore not an incidental implementation detail. It is a direct consequence of the underlying polynomial structure. Once that structure is recognized, the choice of topology follows naturally from the symmetry, or partial symmetry, of the corresponding denominator.

References

1. Weinberg, L. (1975). *Network analysis and synthesis*. Robert E. Krieger Publishing Company Inc., ISBN-10: 0882753215; ISBN-13: 9780882753218
2. Bosse, G. (1951). Siebketten ohne Dämpfungsschwankungen im Durchlaßbereich (Potenzketten). *Frequenz, 5*(10), 279–284. Bibcode: 1951Freq....5..279B.

The Cutoff Frequency 3

Where the constant term and the frequency term meet, the cutoff begins to sparkle.

Chapter 2 established that the denominator of a doubly-terminated low-pass passive ladder network inherits a structural symmetry that reflects the symmetry of the corresponding normalized Butterworth polynomial. That observation identifies the structure that makes exact synthesis possible. The present chapter now interprets the cutoff condition within that structure.

For a doubly-terminated low-pass Butterworth ladder filter, the cutoff frequency is not introduced merely as a target marker on a graph. Rather, it emerges from the relationship between the frequency-independent part of the transfer function and the frequency-dependent terms that shape the transition from passband to stopband. The purpose of this chapter is therefore to explain why the Butterworth -3 dB condition must be referenced to the realized passband level of the doubly-terminated low-pass network and why the balance at cutoff must be understood in Butterworth terms rather than as an isolated numerical coincidence.

3.1 The DC Perspective

At zero frequency, a doubly-terminated low-pass ladder filter reduces to its resistive boundary conditions. The reactive elements do not yet impose the frequency-dependent behavior that characterizes the transition band, and the realized gain is therefore determined by the source and load resistances. For such a network, the DC gain is

K. M. P. Ryan, *Polynomial Symmetry in Butterworth Filter Synthesis*,
Synthesis Lectures on Engineering, Science, and Technology,
https://doi.org/10.1007/978-3-032-33468-8_3

$$G_{\mathrm{DC}} = 20\log_{10}\left(\frac{R_L}{R_S + R_L}\right)\ \mathrm{dB}. \tag{3.1}$$

This quantity establishes the passband reference level. In a Butterworth response, the passband remains maximally flat about this level until the onset of the transition toward cutoff.

The gain at cutoff is measured relative to this established passband value. For a Butterworth low-pass response, the magnitude at cutoff is lower by a factor of $\frac{1}{\sqrt{2}}$, so that

$$G_c = 20\log_{10}\left(\frac{R_L}{R_S + R_L} \cdot \frac{1}{\sqrt{2}}\right)\ \mathrm{dB}. \tag{3.2}$$

Equivalently,

$$G_c = G_{\mathrm{DC}} - 3.0103\,\mathrm{dB}.$$

Thus, both the DC gain and the cutoff gain depend explicitly on the source and load resistances. This is one reason why a doubly-terminated formulation is essential: the passband reference itself is a function of the terminations.

3.2 A Conceptual View of Cutoff

Figure 3.1 is a conceptual sketch of the doubly-terminated Butterworth low-pass response. The passband level is established by the source-load divider, giving the DC gain of Eq. 3.1. The cutoff frequency is then defined relative to that passband reference through Eq. 3.2.

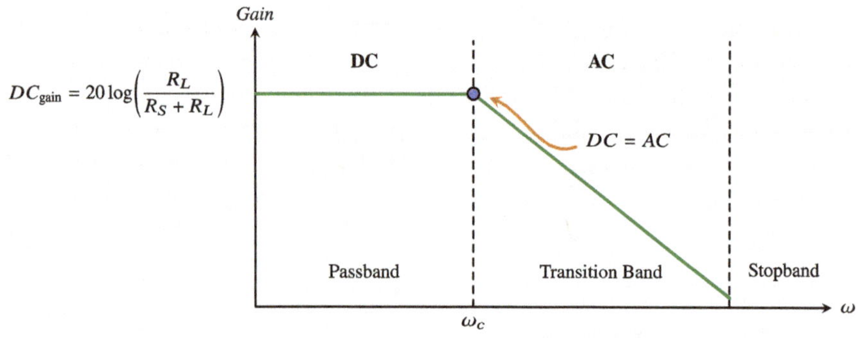

Fig. 3.1 Conceptual sketch of the Butterworth cutoff condition for a doubly terminated low-pass response. The passband reference is set by the source-load divider, and the cutoff point occurs where the frequency-dependent term equals the frequency-independent term in the normalized Butterworth form

The sketch emphasizes the equality condition at cutoff. At ($\omega = \omega_c$), the frequency-independent contribution and the frequency-dependent contribution are equal, so that ($DC_{\text{aspect}} = AC_{\text{aspect}}$). This observation is the starting point for the synthesis approach developed here. The passband level is set by the source-load voltage divider, and the cutoff level is defined relative to that passband level; however, the more important step is recognizing that the cutoff equality must be expressed through the normalized Butterworth polynomial. Once ($DC_{\text{aspect}} = AC_{\text{aspect}}$) is related to the Butterworth coefficient structure, the synthesis problem becomes a matter of matching the corresponding polynomial coefficients to the denominator terms produced by the passive ladder network.

3.3 Where the Frequency-Independent and Frequency-Dependent Terms Meet

The transfer-function denominator of a doubly-terminated low-pass ladder filter contains both a frequency-independent term and terms that depend on frequency. The frequency-independent term establishes the low-frequency reference behavior of the network, while the frequency-dependent terms become increasingly influential as frequency rises through the transition band.

This suggests a useful interpretation of cutoff: the cutoff frequency is the point at which the frequency-dependent contributions become significant in precisely the way required by the Butterworth response. Informally, this is the point where DC meets AC. More precisely, it is the point at which the normalized frequency-independent and frequency-dependent contributions satisfy the Butterworth cutoff condition.

For a normalized Butterworth low-pass response,

$$|H(j\Omega)|^2 = \frac{1}{1 + \Omega^{2n}}, \tag{3.3}$$

where $\Omega = \omega/\omega_c$. At cutoff, $\Omega = 1$, so

$$1 = \Omega^{2n}. \tag{3.4}$$

Equation 3.4 expresses the central balance condition of the normalized Butterworth response: at cutoff, the frequency-independent term and the frequency-dependent term in the magnitude-squared denominator are equal. This is the precise form of the intuition represented in Fig. 3.1.

3.4 Why Balance Alone Is Not Enough

The balancing of frequency-independent and frequency-dependent terms does not, by itself, produce a Butterworth filter. Many transfer functions can exhibit a balance of terms at some frequency without satisfying the Butterworth condition. What distinguishes the Butterworth case is the coefficient structure of the entire denominator.

Accordingly, the design objective is not merely to find a frequency at which one constant term equals one frequency-dependent term. The objective is to place the denominator produced by the ladder network in correspondence with the coefficient structure of the corresponding normalized Butterworth polynomial. Only after that coefficient correspondence has been imposed does the balance at cutoff acquire Butterworth meaning.

This is the point at which Chap. 2 and this chapter meet. Chapter 2 established that the denominator of the passive ladder inherits a symmetry that reflects the symmetry of the Butterworth coefficients. This chapter now shows that, once this correspondence is enforced, the cutoff condition follows from the Butterworth form rather than from approximation.

3.5 The Doubly-Terminated Butterworth Reference Form

For interpretation, it is useful to write the doubly-terminated Butterworth magnitude in the form

$$|H(j\omega)| = \frac{R_L}{R_S + R_L} \cdot \frac{1}{\sqrt{1 + \left(\frac{\omega}{\omega_c}\right)^{2n}}}. \tag{3.5}$$

Equation 3.5 is the Butterworth reference form against which the realized ladder denominator is to be interpreted. At $\omega = 0$, it yields the DC gain of Eq. 3.1. At $\omega = \omega_c$, it yields the cutoff gain of Eq. 3.2. Between these two limits, it describes the maximally flat Butterworth transition for the specified order n.

The importance of Eq. 3.5 is conceptual as much as practical. It makes explicit that the passband level and the cutoff level are both shaped by termination, while the transition shape is governed by the Butterworth denominator. The filter is therefore neither an ideal normalized prototype nor a pure resistor divider; it is the combination of both.

3.6 Transition to the Transformation Rule

Once the denominator of a doubly-terminated low-pass passive ladder transfer function is written in degree-ordered form, its terms may be associated with the corresponding Butterworth coefficients identified in Chap. 2. This association is what turns the cutoff interpretation developed here into a synthesis method.

The resulting procedure is guided by three linked ideas:

1. the realized passband level is established by the source and load resistances,
2. the cutoff condition is defined by the Butterworth relationship between the frequency-independent and frequency-dependent parts of the denominator, and

3. the full Butterworth response is obtained only when the entire denominator is matched coefficient by coefficient.

In this way, the cutoff frequency is not an isolated datum but part of a larger analytical structure. It is one constraint among several, but it is the constraint that ties the realized ladder network to the Butterworth -3 dB condition at the specified frequency.

The next chapter states the transformation rule explicitly. There, the structural symmetry developed in Chap. 2 and the cutoff interpretation developed here are combined into a general degree-matching procedure for exact synthesis.

4 Transformation

Transformation is where structure becomes synthesis.

Chapter 2 identified the structural symmetry that makes exact Butterworth synthesis possible, and Chap. 3 showed how the cutoff condition must be interpreted relative to the realized passband level of a doubly-terminated network. The purpose of the present chapter is to combine those two ideas into an explicit synthesis rule.

The central idea is simple: the denominator of the realized low-pass ladder transfer function is written in degree-ordered form, the normalized Butterworth denominator of the same order is written in coefficient form, and the two are matched term by term. In this monograph, that matching process is called a transformation. It is not a numerical approximation, nor a scaling after the fact. It is the analytical step by which the realized circuit is constrained to satisfy the Butterworth form.

4.1 Projection and Symmetry

The transformation rule depends on the projection described in Chap. 2. The denominator of a doubly-terminated low-pass passive ladder filter is not an arbitrary polynomial. When arranged by degree, its terms exhibit a patterned structure that mirrors the coefficient structure of the normalized Butterworth denominator of the same order.

This correspondence is the reason the synthesis method works. The Butterworth coefficients do not merely describe the target response abstractly; they also provide the multipliers by which the degree-ordered terms of the realized ladder denominator are to be constrained. In this sense, symmetry is not merely descriptive. It is operational.

K. M. P. Ryan, *Polynomial Symmetry in Butterworth Filter Synthesis*,
Synthesis Lectures on Engineering, Science, and Technology,
https://doi.org/10.1007/978-3-032-33468-8_4

For odd-order filters, the projected structure is fully pairable about the center. For even-order filters, the same balancing tendency remains, but the center is unmatched. This distinction affects topology selection, but it does not alter the general form of the transformation rule itself.

4.2 The Realized Ladder Denominator

Consider a doubly-terminated low-pass passive ladder filter of order n. Its denominator may be written in the form

$$D_{\text{ladder}}(\omega) = E_0 + E_1(\omega) + E_2(\omega) + \cdots + E_n(\omega), \tag{4.1}$$

where E_0 is the frequency-independent term and $E_k(\omega)$ is the degree-k frequency-dependent term.

In the earlier informal language of this monograph, E_0 is the DC_{aspect}, while the terms $E_1(\omega), E_2(\omega), \ldots, E_n(\omega)$ are the AC_{aspect} terms. In the remainder of the manuscript, the more precise terminology frequency-independent term and frequency-dependent terms will generally be used.

For the doubly-terminated low-pass ladder networks considered here, the constant term is

$$E_0 = \left(\frac{R_S}{R_L} + 1\right). \tag{4.2}$$

The remaining terms depend on the filter order, the topology, and the element values, but each is associated with a unique degree of ω.

It is useful to write each degree-k term in the form

$$E_k(\omega) = \omega^k T_k, \tag{4.3}$$

where T_k denotes the element combination associated with degree k. The synthesis problem is then to determine the unknown element values embedded in the quantities T_k.

4.3 The Normalized Butterworth Target

The normalized Butterworth denominator of order n may be written as

$$B_n(s) = 1 + a_1 s + a_2 s^2 + \cdots + a_n s^n, \tag{4.4}$$

where the coefficients a_k are the normalized Butterworth coefficients of order n.

As discussed earlier, these coefficients may be generated from Butterworth product factors rather than retrieved from prototype tables. Using the product formula introduced previously,

$$a_k = \prod_{\mu=1}^{k} \frac{\cos((\mu - 1)\gamma)}{\sin(\mu\gamma)}, \qquad \gamma = \frac{\pi}{2n}, \tag{4.5}$$

with $a_0 = 1$.

The importance of the coefficients a_k is twofold. First, they define the target Butterworth response. Second, they provide the exact multipliers used in transforming the realized ladder denominator into Butterworth form.

At a deeper level, these coefficients are themselves assembled from structured Butterworth factors. For lower orders, those factors exhibit simple inversion, scaling, and pairing relationships, which helps explain why the lower-order synthesis chapters remain especially transparent. In higher orders, that simplicity becomes less complete and recursive structure becomes more prominent. The detailed factor relationships are treated later; for the present purpose, it is enough to note that the coefficients used in the transformation are not arbitrary constants, but expressions of an internal structure of their own.

4.4 The Transformation Rule

The transformation used in this monograph is obtained by associating each degree-k term of the realized low-pass ladder denominator with the degree-k coefficient of the normalized Butterworth denominator of the same order. Since the constant term E_0 establishes the realized low-frequency reference of the network, each frequency-dependent term is set equal to that constant term multiplied by its associated Butterworth coefficient.

Accordingly, the general synthesis rule is

$$E_k(\omega_c) = E_0\, a_k, \qquad k = 1, 2, \ldots, n, \tag{4.6}$$

where ω_c is the specified cutoff frequency.

Using Eq. 4.3, this may also be written as

$$\omega_c^k T_k = E_0\, a_k, \qquad k = 1, 2, \ldots, n. \tag{4.7}$$

This is the formal statement of the method that has been expressed informally as

$$AC_{\text{aspect}} = (DC_{\text{aspect}})(\mathit{Butterworth}_{\text{coef}}).$$

The present chapter gives that rule its exact meaning. The transformation is carried out degree by degree, using the constant term of the realized low-pass denominator as the reference quantity and the normalized Butterworth coefficients as the required multipliers.

4.5 Why the Transformation Works

The transformation works because the low-pass denominator of the realized ladder network and the denominator of the normalized Butterworth response are being matched structurally rather than numerically approximated. Each degree of the realized denominator is forced to satisfy the corresponding Butterworth relationship at the specified cutoff frequency.

This does not mean that the realized low-pass ladder denominator becomes identical to the normalized Butterworth denominator as a symbolic expression. Rather, it means that the realized degree-by-degree terms are constrained so that the network reproduces the Butterworth coefficient structure under the specified terminations and cutoff condition.

This is the essential step that turns the synthesis problem into a solvable system. Once Eq. 4.6 is imposed, each degree contributes one equation. The unknowns are the reactive element values of the ladder network.

4.6 Number of Equations and Unknowns

For an nth-order ladder filter, there are n reactive element values to be determined. The transformation produces one degree-matching equation for each of the n nonconstant degrees. Thus, the method yields

$$\text{number of equations} = \text{number of unknowns} = n.$$

This is one of the most useful features of the approach. The order of the filter determines not only the degree of the Butterworth target polynomial, but also the number of synthesis constraints and the number of unknown element values. The resulting system may admit one solution or multiple solutions, depending on the order, topology, and terminations. The structure of that solution set is the subject of the next chapter.

4.7 Topology and Applicability

The transformation rule itself is general across the low-pass ladder forms considered in this monograph, but its implementation depends on the topology chosen for the realized network.

For odd-order filters, either shunt-first or series-first topologies may be used because the projected denominator structure remains fully symmetric. For even-order filters, topology must be selected in accordance with the relative magnitudes of R_S and R_L, since the unmatched center structure imposes an implementation constraint. In that case, the transformation is still applied degree by degree, but only after the appropriate topology has been chosen.

Thus, topology selection does not replace the transformation; it precedes it.

4.8 Interpretation

The transformation developed here should be understood as the analytical heart of the synthesis procedure. It takes the qualitative ideas developed earlier in the monograph and turns them into design equations.

Chapter 2 established that symmetry provides the structural basis for coefficient matching. Chapter 3 established that the cutoff condition must be interpreted in Butterworth terms rather than merely as a balance between constant and frequency-dependent contributions. The present chapter combines those two ideas into a usable rule:

> each degree-k frequency-dependent term of the realized ladder denominator is set equal, at cutoff, to the frequency-independent term multiplied by the corresponding normalized Butterworth coefficient.

Everything that follows in the order-specific synthesis chapters is an application of that rule.

4.9 Transition to Solution Structure

The purpose of this chapter has been to state the general rule by which structural symmetry and Butterworth cutoff interpretation are converted into synthesis equations. Once that rule is imposed, however, a new question arises: what kind of solution set does the resulting system produce?

For an nth-order filter, the degree-matching procedure yields n equations in n unknown reactive element values. That system may produce one solution, multiple real solutions, or a mixture of real and non-real candidates, depending on order, topology, and terminations.

Accordingly, Chap. 5 examines the solution structure generated by the transformation rule before the monograph proceeds to the order-specific synthesis chapters.

Solution Structure 5

A synthesis method is judged not only by whether it yields a solution, but by what kind of solution it yields.

Chapter 4 established the transformation rule used throughout this monograph: each degree-k frequency-dependent term of the realized ladder denominator is set equal, at cutoff, to the frequency-independent term multiplied by the corresponding normalized Butterworth coefficient. Once that rule is imposed, the synthesis problem becomes a system of algebraic equations having unknown reactive element values.

The present chapter examines the structure of that system. The important question is not merely whether the method produces a solution, but what kind of solution it produces. For a given filter order and topology, the system may yield one solution, several solutions, or a mixture of real and non-real candidates. The purpose of this chapter is to explain that behavior before the order-specific synthesis chapters begin.

5.1 Number of Equations and Unknowns

For an nth-order doubly-terminated passive ladder filter, there are n reactive elements to be determined. These may be capacitors, inductors, or alternating sequences of both, depending on topology and filter type. The transformation rule developed in Chap. 4 yields one matching equation for each nonconstant degree of the denominator. Accordingly,

$$\text{number of equations} = \text{number of unknowns} = n.$$

K. M. P. Ryan, *Polynomial Symmetry in Butterworth Filter Synthesis*,
Synthesis Lectures on Engineering, Science, and Technology,
https://doi.org/10.1007/978-3-032-33468-8_5

This one-to-one correspondence is one of the central structural features of the method. The order of the filter determines the number of degrees in the denominator, the number of Butterworth coefficients to be matched, and the number of unknown element values to be solved.

Thus, first-order synthesis yields one equation in one unknown, second-order synthesis yields two equations in two unknowns, and so on. As the order increases, the size of the algebraic system increases accordingly, and with it the richness of the solution set.

5.2 Why Multiple Solutions Can Occur

Although an nth-order synthesis problem contains n equations in n unknowns, this does not imply that there is only one solution. Algebraic systems of this kind may admit multiple candidate solutions, all satisfying the same cutoff frequency and termination requirements.

This is not a defect of the method; it is a property of the design space. Different combinations of reactive element values may produce denominators that satisfy the same Butterworth coefficient constraints. When this occurs, the synthesis method reveals that the design target is realizable in more than one exact way.

This behavior becomes especially important beginning with second order, where more than one valid real solution may satisfy the same design requirements. In higher orders, the number of candidate solutions may increase further, and the distinction between algebraic possibility and physical realizability becomes correspondingly more important.

5.3 Real and Non-real Solutions

Not every algebraic solution produced by the transformed system is physically acceptable as a passive filter realization. In the present work, a distinction is made between real solutions and non-real solutions.

A real solution is one in which every solved element value is real-valued and therefore corresponds directly to a realizable passive component value under the assumptions of the synthesis. These are the solutions retained as valid filter designs throughout this monograph.

A non-real solution is one in which one or more solved element values contain a nonzero imaginary component. Such solutions may still satisfy the transformed algebraic system formally, but they do not correspond directly to passive capacitor and inductor values in the intended physical sense. Accordingly, they are not retained as valid realizations in the present development.

This distinction matters because the synthesis method is exact at the algebraic level, but the design objective is a realizable passive ladder network. Algebraic admissibility and physical admissibility are therefore related, but not identical.

5.4 Topology and Realizability

Topology plays a direct role in determining whether a given candidate solution is physically meaningful. As established earlier, odd-order filters permit either shunt-first or series-first realization, whereas even-order filters require topology selection consistent with the relative magnitudes of R_S and R_L.

This means that realizability is constrained in two ways. First, the solved element values must be real. Second, the topology under which those values are interpreted must be appropriate for the termination conditions. A candidate solution that is algebraically valid in one formal arrangement may fail to produce a meaningful passive realization when paired with an inappropriate topology.

For this reason, the synthesis procedure used in the later chapters always proceeds in the following order:

1. select the appropriate topology,
2. write the realized ladder denominator for that topology,
3. apply the transformation rule,
4. solve the resulting system, and
5. retain only the real solutions that correspond to valid passive element values.

5.5 Single-solution and Multi-solution Cases

The simplest case occurs at first order. There is one unknown reactive element and one equation, so the solution structure is correspondingly simple. In that setting, the method yields a single exact element value for the specified cutoff frequency and terminations.

Beginning at second order, the situation becomes more interesting. The algebraic system may yield more than one real solution, meaning that more than one passive ladder realization satisfies the same Butterworth requirements. This is an important feature of the method, since it reveals that the synthesis target need not determine a unique network.

At still higher orders, the system may yield a mixture of real and non-real solutions. The appearance of such solution sets is not surprising from the algebraic standpoint, since higher-order systems naturally admit richer root structures. The role of the synthesis procedure is therefore not merely to solve, but to classify.

5.6 Interpreting Multiple Real Solutions

When more than one real solution exists, each real solution corresponds to a distinct exact realization of the same Butterworth design target. These realizations share the same specified cutoff frequency, the same source and load terminations, and the same Butterworth coefficient-matching conditions, but they differ in the distribution of element values within the ladder.

This is an important point. The existence of multiple real solutions does not imply ambiguity in the target response. Rather, it implies non-uniqueness in the exact passive realization of that response.

From a practical standpoint, multiple real solutions may be useful. One exact solution may be preferable to another because of element spread, available component values, sensitivity, physical size, or implementation convenience. The synthesis method developed here therefore provides not only exact element values but, in some cases, a choice among exact realizations.

5.7 Why Non-real Solutions Are Excluded

In the examples developed in this monograph, solutions containing nonzero imaginary components are not retained as valid passive filter designs. The reason is straightforward: the intended realizations are passive ladder networks composed of ordinary capacitors and inductors having real-valued element magnitudes.

A non-real solution may still be mathematically informative. It may reveal something about the structure of the transformed system, the constraints imposed by the selected topology, or the boundary between realizable and non-realizable designs. Even so, the design chapters that follow focus on real passive realizations and therefore treat non-real solutions as inadmissible for practical synthesis.

This convention is consistent throughout the manuscript: exact synthesis is pursued, but only within the class of real passive ladder networks under the stated termination conditions.

5.8 Relation to Higher Order

As filter order increases, the solution structure generally becomes richer. More unknowns produce more complicated algebraic systems, and these systems may generate larger collections of candidate solutions. In lower orders, much of this structure can still be understood transparently. In higher orders, additional algebraic complexity emerges and recursive structure becomes more prominent.

For this reason, the synthesis chapters that follow proceed order by order. Each chapter applies the same transformation principle, but the resulting solution structure is examined in the specific context of that order. In this way, the reader can see how the method scales while maintaining a clear interpretation of what each solution means.

5.9 Transition to the Synthesis Chapters

The purpose of this chapter has been to explain the solution structure produced by the transformation rule before turning to specific filter orders. The remaining chapters do not introduce a new method at each order; rather, they apply the same degree-matching principle to progressively richer solution spaces.

Accordingly, Chap. 6 begins with the simplest possible case: the first-order doubly-terminated passive low-pass Butterworth ladder. There, one frequency-dependent degree and one unknown reactive element produce the first exact worked realization of the method.

Part II
Doubly-Terminated Passive Butterworth Filter Network Synthesis

Abstract This chapter presents the exact synthesis of first-order doubly-terminated passive Butterworth ladder filters. Because first-order Butterworth networks are odd-order and therefore structurally symmetric, either shunt-first or series-first low-pass topologies may be used regardless of the relative values of R_S and R_L. The chapter derives the corresponding first-order synthesis equations, develops the associated time-constant interpretation, and then shows how the resulting low-pass networks may be converted directly into high-pass networks by reactive duality and frequency inversion. Although the examples are first-order, the chapter serves as the simplest concrete application of the transformation rule and solution structure developed earlier in the monograph.

1st-Order Doubly-Terminated Passive Butterworth Filter Network Synthesis 6

First in a long ordered line.

Chapters 4 and 5 established the two principles that govern the synthesis chapters that follow. First, each degree-k frequency-dependent term of the realized low-pass ladder denominator is set equal, at cutoff, to the frequency-independent term multiplied by the corresponding normalized Butterworth coefficient. Second, the resulting algebraic system is to be interpreted in terms of its real and non-real solution structure. The present chapter applies those principles in the simplest possible case: the first-order doubly-terminated low-pass passive Butterworth ladder.

Unlike traditional first-order treatments, the present approach integrates exact component calculation with explicit source and load terminations and demonstrates a method that generalizes naturally to higher-order filters. The purpose of the chapter is therefore twofold. First, it shows how the transformation rule is applied in the least complicated setting. Second, it establishes patterns of interpretation that recur throughout the remainder of the monograph: degree-matching, time-constant interpretation, topology selection, and low-pass to high-pass conversion.

Although the examples here are first-order, the principle is fully general. The first-order case is simply the most elementary instance of the broader synthesis framework.

K. M. P. Ryan, *Polynomial Symmetry in Butterworth Filter Synthesis*,
Synthesis Lectures on Engineering, Science, and Technology,
https://doi.org/10.1007/978-3-032-33468-8_6

6.1 Odd Symmetry

A first-order doubly-terminated low-pass passive Butterworth filter is odd-order and therefore structurally symmetric. Consequently, either a shunt-first or a series-first circuit topology may be used regardless of the relative values of R_S and R_L.

This is the simplest case of the odd-order flexibility discussed earlier. Because the first-order low-pass denominator contains only one frequency-dependent term and one frequency-independent term, the synthesis problem reduces to one matching equation having one unknown.

6.2 Low-Pass Shunt-First

The transfer function for the shunt-first first-order doubly-terminated low-pass filter of Fig. 6.1 is

$$\frac{V_{OUT}}{V_{IN}} = \frac{1}{\omega C_1 R_S + \left(\frac{R_S}{R_L} + 1\right)}. \tag{6.1}$$

The corresponding normalized Butterworth transfer function, written in product-factor form, is (Fig. 6.2)

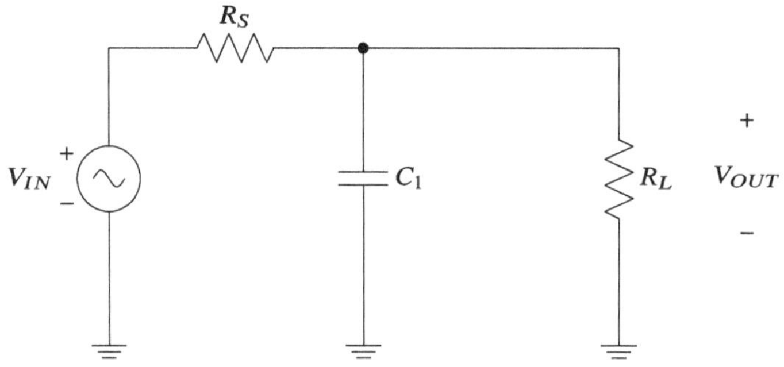

Fig. 6.1 First-order shunt-first low-pass filter

$$\frac{V_{OUT}}{V_{IN}} = \frac{1}{\omega C_1 R_S + \left(\frac{R_S}{R_L} + 1\right)}$$

(a) Transfer function of Fig. 6.1

$$\frac{V_{OUT}}{V_{IN}} = \frac{1}{\left(\frac{\cos(0)}{\sin(\frac{1}{2}\pi)}\right) s + 1}$$

(b) Corresponding normalized Butterworth form using Eq. 2.1

Fig. 6.2 First-order shunt-first low-pass equations

$$\frac{V_{OUT}}{V_{IN}} = \frac{1}{\left(\frac{\cos(0)}{\sin(\frac{1}{2}\pi)}\right)s + 1}. \tag{6.2}$$

For a first-order filter there is one frequency-dependent degree, one unknown element value, and therefore one synthesis equation. Applying the transformation rule developed in Chap. 4,

$$AC_{aspect} = (DC_{aspect})(Butterworth_{coef}), \tag{6.3}$$

gives the exact first-order matching condition.

For the shunt-first case, the frequency-dependent term is $\omega C_1 R_S$, while the frequency-independent term is $\left(\frac{R_S}{R_L} + 1\right)$. The associated Butterworth coefficient is

$$\left(\frac{\cos(0)}{\sin\left(\frac{1}{2}\pi\right)}\right).$$

Applying the transformation rule gives

$$\omega C_1 R_S = \left(\frac{R_S}{R_L} + 1\right)\left(\frac{\cos(0)}{\sin\left(\frac{1}{2}\pi\right)}\right), \tag{6.4}$$

which simplifies to

$$\omega C_1 R_S = \left(\frac{R_S}{R_L} + 1\right). \tag{6.5}$$

Solving for C_1 yields

$$C_1 = \frac{1}{\omega_c\left(\frac{R_S R_L}{R_S + R_L}\right)}. \tag{6.6}$$

Equation 6.6 gives the exact capacitor value required for a first-order shunt-first doubly-terminated low-pass Butterworth filter having specified R_S, R_L, and ω_c.

Equivalently, if the network already exists and the element value is known, the cutoff frequency may be written as

$$\omega_c = \frac{1}{C_1\left(\frac{R_S R_L}{R_S + R_L}\right)} = \frac{1}{time}. \tag{6.7}$$

6.2.1 Time Constant

The capacitive time constant is

$$\tau_C = C_1\left(\frac{R_S R_L}{R_S + R_L}\right). \tag{6.8}$$

Thus, the capacitive time constant in a doubly-terminated first-order shunt-first low-pass filter is the capacitor multiplied by the parallel combination of source and load resistance.

Using Eq. 6.8, the first-order relationship may be written as

$$\omega\tau_C = \sin\left(\frac{1}{2}\pi\right) = \frac{\cos(0)}{\sin(\frac{1}{2}\pi)} = \csc\left(\frac{1}{2}\pi\right). \tag{6.9}$$

This is the first appearance of the trigonometric time-constant identities that later re-emerge in higher-order synthesis and are collected in Chap. 12.

6.2.2 Design Example—First-Order Low-Pass Shunt-First

The design requirements are:

- $R_S = 203\,\Omega$
- $R_L = 17\,\Omega$
- $f_c = 5\,\text{kHz}$
- $\omega_c = 31.416 \times 10^3\,\text{rad/s}$.

The gain at DC, or throughout the passband as established by the terminations, is

$$20\log_{10}\left(\frac{17\,\Omega}{203\,\Omega + 17\,\Omega}\right) = -22.23947\,\text{dB}.$$

Therefore the gain at cutoff is

$$20\log_{10}\left(\frac{1}{\sqrt{2}} * \frac{17\,\Omega}{203\,\Omega + 17\,\Omega}\right) = -25.24977\,\text{dB},$$

or

$$-22.23947\,\text{dB} - 3.010299\,\text{dB} = -25.24977\,\text{dB}.$$

The resulting solution is (Fig. 6.3)

$$C_1 = \frac{1}{\omega_c\left(\frac{R_S R_L}{R_S + R_L}\right)} = \frac{1}{31.416 \times 10^3\left(\frac{(203)(17)}{203+17}\right)} = 2.029213\,\mu\text{F}.$$

6.3 Low-Pass Series-First

The transfer function for the series-first first-order doubly-terminated low-pass filter of Fig. 6.4 is

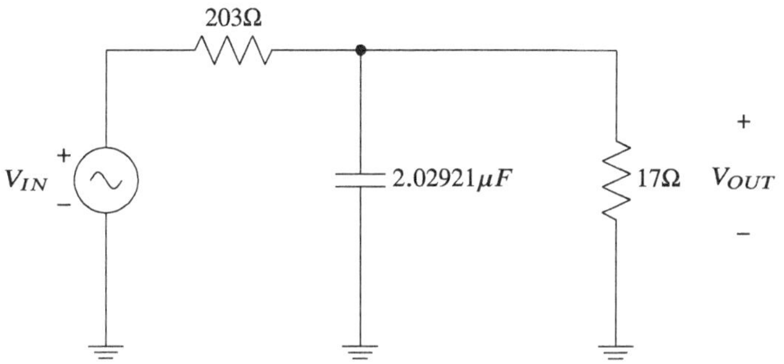

Fig. 6.3 First-order shunt-first low-pass filter design

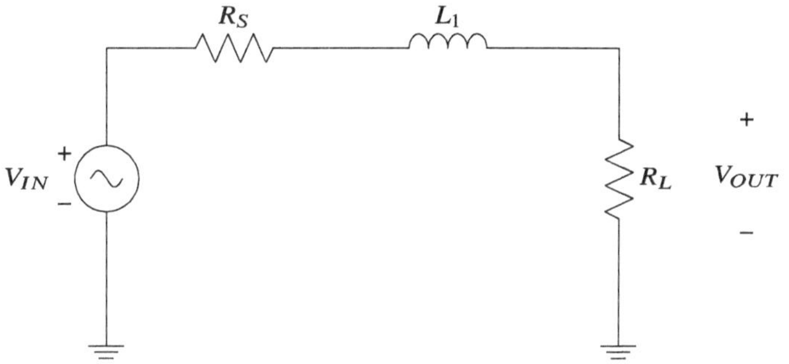

Fig. 6.4 First-order series-first low-pass filter

$$\frac{V_{OUT}}{V_{IN}} = \frac{1}{\omega \frac{L_1}{R_L} + \left(\frac{R_S}{R_L} + 1\right)}$$

(a) Transfer function of Fig. 6.4

$$\frac{V_{OUT}}{V_{IN}} = \frac{1}{\left(\frac{\cos(0)}{\sin(\frac{1}{2}\pi)}\right) s + 1}$$

(b) Corresponding normalized Butterworth form using Eq. 2.1

Fig. 6.5 First-order series-first low-pass equations

$$\frac{V_{OUT}}{V_{IN}} = \frac{1}{\omega \frac{L_1}{R_L} + \left(\frac{R_S}{R_L} + 1\right)}. \tag{6.10}$$

The corresponding normalized Butterworth transfer function is again (Fig. 6.5)

$$\frac{V_{OUT}}{V_{IN}} = \frac{1}{\left(\frac{\cos(0)}{\sin(\frac{1}{2}\pi)}\right) s + 1}. \tag{6.11}$$

Again, only the denominators are relevant for synthesis. Applying the transformation rule,

$$AC_{aspect} = (DC_{aspect})(Butterworth_{coef}), \tag{6.12}$$

gives

$$\omega \frac{L_1}{R_L} = \left(\frac{R_S}{R_L} + 1\right)\left(\frac{\cos(0)}{\sin\left(\frac{1}{2}\pi\right)}\right), \tag{6.13}$$

which simplifies to

$$\omega \frac{L_1}{R_L} = \left(\frac{R_S}{R_L} + 1\right). \tag{6.14}$$

Solving for L_1 yields

$$L_1 = \frac{1}{\omega_c \left(\frac{1}{R_S + R_L}\right)}. \tag{6.15}$$

Thus, the exact inductor value for a first-order series-first doubly-terminated low-pass Butterworth filter is determined directly from the specified terminations and cutoff frequency.

Equivalently, the cutoff frequency of an existing network may be written as

$$\omega_c = \frac{1}{L_1 \left(\frac{1}{R_S + R_L}\right)} = \frac{1}{time}. \tag{6.16}$$

6.3.1 Time Constant

The inductive time constant is

$$\tau_L = L_1 \left(\frac{1}{R_S + R_L}\right). \tag{6.17}$$

Thus, the inductive time constant in a doubly-terminated first-order series-first low-pass filter is the inductor divided by the series combination of source and load resistance.

Using Eqs. 6.8 and 6.17, the first-order low-pass relationships may be written as

$$\omega \tau_C = \omega \tau_L = \sin\left(\frac{1}{2}\pi\right) = \frac{\cos(0)}{\sin(\frac{1}{2}\pi)} = \csc\left(\frac{1}{2}\pi\right). \tag{6.18}$$

6.3.2 Design Example—First-Order Low-Pass Series-First

The design requirements are:

- $R_S = 10.127\,\text{k}\Omega$
- $R_L = 123.456\,\text{k}\Omega$

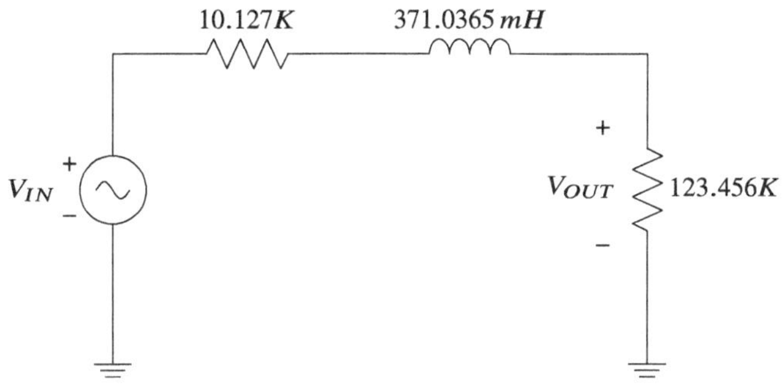

Fig. 6.6 First-order series-first low-pass filter design

- $f_c = 57.3\,\text{kHz}$
- $\omega_c = 360.0265 \times 10^3\,\text{rad/s}$.

The gain at DC, or throughout the passband as established by the terminations, is

$$20\log_{10}\left(\frac{123.456\,\text{k}\Omega}{10.127\,\text{k}\Omega + 123.456\,\text{k}\Omega}\right) = -0.68478\,\text{dB}.$$

Therefore the gain at cutoff is

$$20\log_{10}\left(\frac{1}{\sqrt{2}} * \frac{123.456\,\text{k}\Omega}{10.127\,\text{k}\Omega + 123.456\,\text{k}\Omega}\right) = -3.695079\,\text{dB},$$

or

$$-0.68478\,\text{dB} - 3.010299\,\text{dB} = -3.695079\,\text{dB}.$$

The resulting solution is (Fig. 6.6)

$$L_1 = \frac{1}{\omega_c\left(\frac{1}{R_S+R_L}\right)} = \frac{1}{360.0265 \times 10^3\left(\frac{1}{10.127\times10^3+123.456\times10^3}\right)} = 371.0365\,\text{mH}.$$

6.4 High-Pass Conversion

A low-pass to high-pass conversion may be accomplished directly through frequency inversion and reactive duality. In this approach, a capacitor in the low-pass ladder is replaced by an inductor in the corresponding high-pass ladder, and an inductor in the low-pass ladder is replaced by a capacitor. The termination resistances do not need to be re-derived during this conversion step; they have already been accounted for in the low-pass synthesis.

This conversion preserves the Butterworth coefficient structure and the cutoff frequency of the original low-pass design while producing the corresponding high-pass response. More generally, the same conversion principle applies to every order considered in this monograph.

For a shunt capacitor in the low-pass network, the corresponding high-pass shunt inductor is obtained by equating impedances:

$$\omega L_{HP} = \frac{1}{\omega C_{LP}}.$$

Solving for the high-pass inductance gives

$$L_{HP} = \frac{1}{\omega^2 C_{LP}}. \tag{6.19}$$

Similarly, for a series inductor in the low-pass network, the corresponding high-pass series capacitor is obtained from

$$\frac{1}{\omega C_{HP}} = \omega L_{LP},$$

which yields

$$C_{HP} = \frac{1}{\omega^2 L_{LP}}. \tag{6.20}$$

Equations 6.19 and 6.20 may be applied element by element to convert any low-pass ladder into the corresponding high-pass ladder of the same order.

6.5 High-Pass Shunt-First

A first-order shunt-first high-pass filter is obtained by first synthesizing the corresponding first-order shunt-first low-pass ladder and then converting the low-pass shunt capacitor using Eq. 6.19. Therefore, the high-pass example is a transformation example that uses the shunt-first low-pass result to obtain a low-pass precursor, which is then converted into the final high-pass ladder. The low-pass shunt capacitor becomes the high-pass shunt inductor.

For the shunt-first low-pass precursor, Eq. 6.19 gives

$$L_{HP} = \frac{1}{\omega^2 C_{LP}}.$$

The resulting network is a first-order shunt-first doubly-terminated high-pass Butterworth filter (Fig. 6.7).

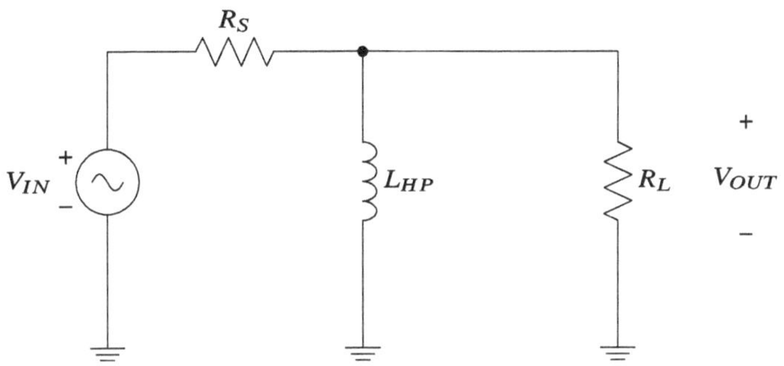

Fig. 6.7 First-order shunt-first high-pass filter

6.5.1 Design Example—First-Order High-Pass Shunt-First

The design requirements are:

- $R_S = 127.9\,\text{k}\Omega$
- $R_L = 536.8\,\text{k}\Omega$
- $f_c = 36.75\,\text{kHz}$
- $\omega_c = 230.907 \times 10^3\,\text{rad/s}$.

The passband gain established by the terminations is

$$20\log_{10}\left(\frac{536.8\,\text{k}\Omega}{127.9\,\text{k}\Omega + 536.8\,\text{k}\Omega}\right) = -1.856263\,\text{dB}.$$

Therefore the gain at the high-pass cut-on frequency is

$$20\log_{10}\left(\frac{1}{\sqrt{2}} * \frac{536.8\,\text{k}\Omega}{127.9\,\text{k}\Omega + 536.8\,\text{k}\Omega}\right) = -4.866562\,\text{dB},$$

or

$$-1.856263\,\text{dB} - 3.010299\,\text{dB} = -4.866562\,\text{dB}.$$

First, the corresponding low-pass capacitor is synthesized (Fig. 6.8):

$$C_1 = \frac{1}{\omega_c\left(\frac{R_S R_L}{R_S + R_L}\right)} = \frac{1}{230.907 \times 10^3\left(\frac{(127.9\times10^3)(536.8\times10^3)}{127.9\times10^3+536.8\times10^3}\right)} = 41.92812\,\text{pF}.$$

Applying Eq. 6.19 gives (Fig. 6.9)

$$L_{HP} = \frac{1}{\omega_c^2 C_{LP}} = \frac{1}{(230.907 \times 10^3)^2(41.92812\,\text{pF})} = 447.3221\,\text{mH}.$$

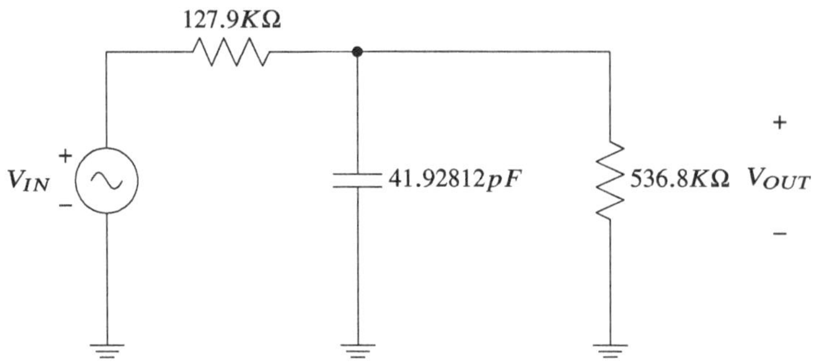

Fig. 6.8 First-order shunt-first low-pass filter used for high-pass conversion

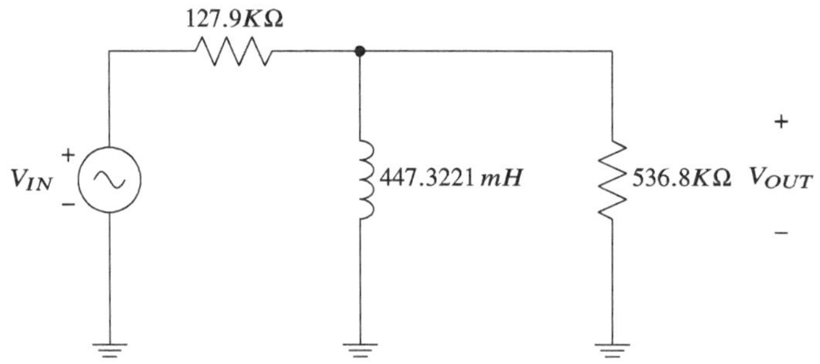

Fig. 6.9 First-order shunt-first high-pass filter design

6.6 High-Pass Series-First

A first-order series-first high-pass filter is obtained by first synthesizing the corresponding first-order series-first low-pass ladder and then converting the low-pass series inductor using Eq. 6.20. Therefore, the high-pass example is a transformation example that uses the series-first low-pass result to obtain a low-pass precursor, which is then converted into the final high-pass ladder. The low-pass series inductor becomes the high-pass series capacitor.

For the series-first low-pass precursor, Eq. 6.20 gives

$$C_{HP} = \frac{1}{\omega^2 L_{LP}}.$$

The resulting network is a first-order series-first doubly-terminated high-pass Butterworth filter (Fig. 6.10).

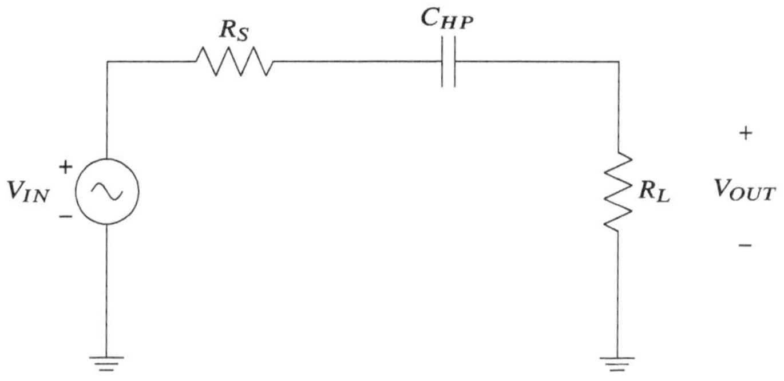

Fig. 6.10 First-order series-first high-pass filter

6.6.1 Design Example—First-Order High-Pass Series-First

The design requirements are:

- $R_S = 100\,\Omega$
- $R_L = 101\,\Omega$
- $f_c = 2.54\,\text{kHz}$
- $\omega_c = 15.95929 \times 10^3\,\text{rad/s}$.

The passband gain established by the terminations is

$$20\log_{10}\left(\frac{101\,\Omega}{100\,\Omega + 101\,\Omega}\right) = -5.977493\,\text{dB}.$$

Therefore the gain at the high-pass cut-on frequency is

$$20\log_{10}\left(\frac{1}{\sqrt{2}} * \frac{101\,\Omega}{100\,\Omega + 101\,\Omega}\right) = -8.987792\,\text{dB},$$

or

$$-5.977493\,\text{dB} - 3.010299\,\text{dB} = -8.987792\,\text{dB}.$$

First, the corresponding low-pass inductance is synthesized (Fig. 6.11):

$$L_1 = \frac{1}{\omega_c\left(\frac{1}{R_S+R_L}\right)} = \frac{1}{15.95929 \times 10^3\left(\frac{1}{100+101}\right)} = 12.59454\,\text{mH}.$$

Applying Eq. 6.20 gives (Fig. 6.12)

$$C_{HP} = \frac{1}{\omega_c^2 L_{LP}} = \frac{1}{(15.95929 \times 10^3)^2(12.59454\,\text{mH})} = 311.7385\,\text{nF}.$$

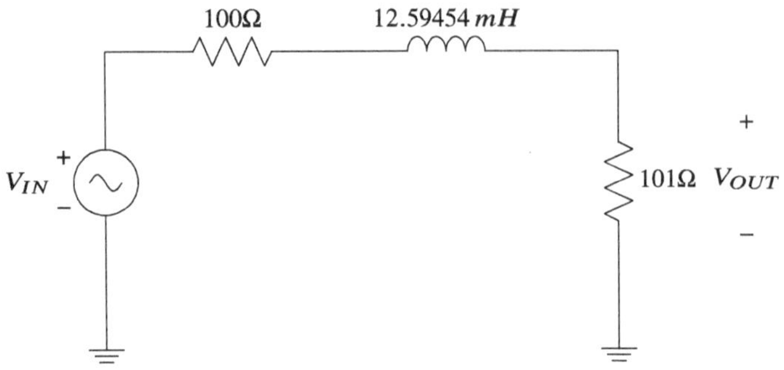

Fig. 6.11 First-order series-first low-pass filter used for high-pass conversion

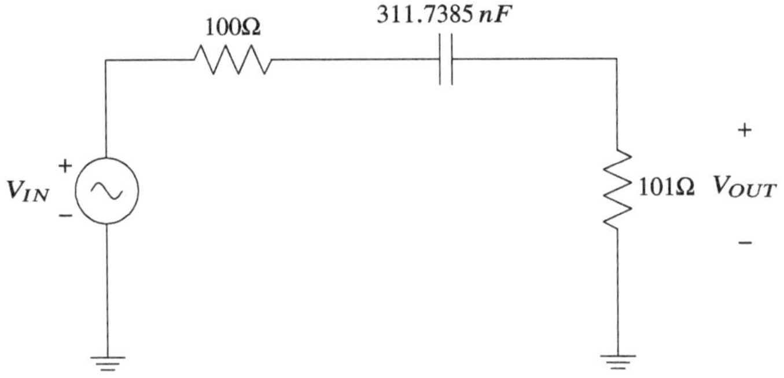

Fig. 6.12 First-order series-first high-pass filter design

6.7 Conclusion

The first-order case is the simplest exact application of the synthesis framework developed in the preceding chapters. Because there is only one frequency-dependent degree and one unknown reactive element, the transformation rule reduces the design problem to a single exact matching equation.

Even in this simplest case, the essential features of the broader method are already visible: explicit treatment of source and load terminations, degree-by-degree matching to the Butterworth target, interpretation through time constants, and direct low-pass to high-pass conversion. These features remain in the higher-order chapters, but their solution structure becomes richer.

Accordingly, Chap. 7 turns to the second-order case, where exact synthesis first yields multiple real passive realizations for the same Butterworth design target.

7 2nd-Order Doubly-Terminated Passive Butterworth Filter Network Synthesis

Second order is where exact synthesis first becomes plural.

Chapter 6 presented the first exact worked realization of the synthesis method in the simplest possible case: one degree-matching equation in one unknown reactive element value. The present chapter is the next important step. It is the first point in the monograph at which exact synthesis yields more than one passive realization for the same Butterworth design target.

As established in Chaps. 4 and 5, the method remains the same in principle: each degree-k frequency-dependent term of the realized ladder denominator is set equal, at cutoff, to the frequency-independent term multiplied by the corresponding normalized Butterworth coefficient. What changes at second order is the structure of the solution set. This is therefore the first chapter in which exact synthesis becomes visibly non-unique.

7.1 Even Symmetry

Second-order doubly-terminated low-pass passive Butterworth filters are even-order and therefore structurally asymmetrical in the sense described earlier. Accordingly, a shunt-first circuit topology must be used when $R_S > R_L$, and a series-first circuit topology must be used when $R_S < R_L$. If $R_S = R_L$, either shunt-first or series-first topology may be used.

For the second-order case, the synthesis procedure results in two candidate circuit solutions. When those solutions are real, both are valid exact realizations of the specified Butterworth design target. Solutions containing imaginary components are not retained as valid passive designs in the present work.

K. M. P. Ryan, *Polynomial Symmetry in Butterworth Filter Synthesis*,
Synthesis Lectures on Engineering, Science, and Technology,
https://doi.org/10.1007/978-3-032-33468-8_7

7.2 Low-Pass Shunt-First

The transfer function for the second-order shunt-first doubly-terminated low-pass filter in Fig. 7.1 is

$$\frac{V_{OUT}}{V_{IN}} = \frac{1}{\omega^2 C_1 L_2 \frac{R_S}{R_L} + \omega\left(C_1 R_S + \frac{L_2}{R_L}\right) + \left(\frac{R_S}{R_L} + 1\right).} \tag{7.1}$$

The corresponding normalized Butterworth transfer function, written in product-factor form, is

$$\frac{V_{OUT}}{V_{IN}} = \frac{1}{s^2\left(\left(\frac{\cos(0)}{\sin(\frac{1}{4}\pi)}\right)\left(\frac{\cos(\frac{1}{4}\pi)}{\sin(\frac{2}{4}\pi)}\right)\right) + s\left(\frac{\cos(0)}{\sin(\frac{1}{4}\pi)}\right) + 1.} \tag{7.2}$$

The second-degree term of the circuit denominator,

$$\omega^2 C_1 L_2 \frac{R_S}{R_L},$$

is associated with the second-degree Butterworth term

$$s^2\left(\left(\frac{\cos(0)}{\sin(\frac{1}{4}\pi)}\right)\left(\frac{\cos(\frac{1}{4}\pi)}{\sin(\frac{2}{4}\pi)}\right)\right).$$

The first-degree term of the circuit denominator,

$$\omega\left(C_1 R_S + \frac{L_2}{R_L}\right),$$

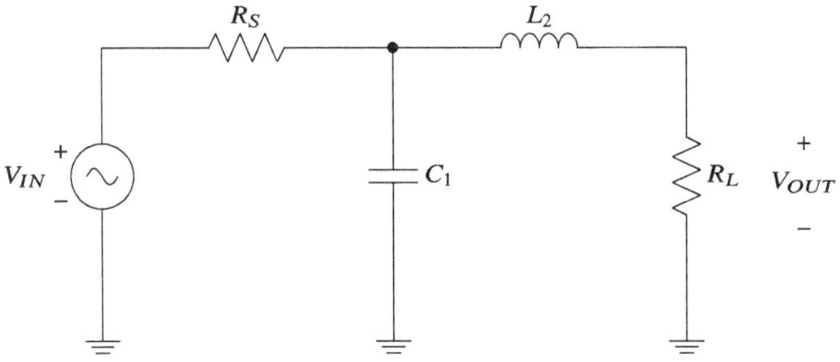

Fig. 7.1 Second-order shunt-first doubly-terminated low-pass filter for $R_S > R_L$

is associated with the first-degree Butterworth term

$$s\left(\frac{\cos(0)}{\sin(\frac{1}{4}\pi)}\right).$$

Applying the transformation rule gives

$$\begin{aligned}AC_{aspect} &= \left(DC_{aspect}\right)\left(Butterworth_{coef}\right)\\ \omega^2 C_1 L_2 \frac{R_S}{R_L} &= \left(\frac{R_S}{R_L}+1\right)\left(\left(\frac{\cos(0)}{\sin(\frac{1}{4}\pi)}\right)\left(\frac{\cos(\frac{1}{4}\pi)}{\sin(\frac{2}{4}\pi)}\right)\right)\\ \omega\left(C_1 R_S + \frac{L_2}{R_L}\right) &= \left(\frac{R_S}{R_L}+1\right)\left(\frac{\cos(0)}{\sin(\frac{1}{4}\pi)}\right)\end{aligned} \tag{7.3}$$

This system simplifies to

$$\omega^2 C_1 L_2 \frac{R_S}{R_L} = \left(\frac{R_S}{R_L}+1\right), \tag{7.4}$$

and

$$\omega\left(C_1 R_S + \frac{L_2}{R_L}\right) = \left(\frac{R_S}{R_L}+1\right)(\sqrt{2}) = \left(\frac{R_S}{R_L}+1\right)\left(\csc\left(\frac{\pi}{4}\right)\right). \tag{7.5}$$

Since there are two unknowns, these may be expressed independently using quadratic closed forms. The inductor values are given by

$$L = \frac{\sin(\frac{1}{4}\pi)}{\omega}\left((R_S+R_L) \pm \sqrt{|R_S^2 - R_L^2|}\right), \tag{7.6}$$

and the capacitor values by

$$C = \frac{\sin(\frac{1}{4}\pi)}{\omega}\left(\left(\frac{1}{R_S}+\frac{1}{R_L}\right) \pm \sqrt{\left|\frac{1}{R_S^2} - \frac{1}{R_L^2}\right|}\right). \tag{7.7}$$

There is a pairing relation between the two L values and the two C values. When the $+$ sign is used in Eq. 7.6, the $-$ sign must be used in Eq. 7.7 to obtain the corresponding capacitor value. Conversely, when the $-$ sign is used in Eq. 7.6, the $+$ sign is used in Eq. 7.7.

Thus, the second-order shunt-first synthesis problem may produce two distinct exact real circuit solutions.

7.2.1 Design Example—Second-Order Low-Pass Shunt-First Using the Quadratic Forms

The design requirements are:

- $R_S = 20.3\,\mathrm{k\Omega}$
- $R_L = 17.8\,\mathrm{k\Omega}$
- $f_c = 105\,\mathrm{kHz}$
- $\omega_c = 659.7344 \times 10^3\,\mathrm{rad/s}$.

The gain at DC, or throughout the passband as established by the terminations, is

$$20\log_{10}\left(\frac{17.8\,\mathrm{k\Omega}}{20.3\,\mathrm{k\Omega} + 17.8\,\mathrm{k\Omega}}\right) = -6.610099\,\mathrm{dB}.$$

Therefore the gain at cutoff is

$$20\log_{10}\left(\frac{1}{\sqrt{2}} * \frac{17.8\,\mathrm{k\Omega}}{17.8\,\mathrm{k\Omega} + 20.3\,\mathrm{k\Omega}}\right) = -9.620399\,\mathrm{dB},$$

or

$$-6.610099\,\mathrm{dB} - 3.010299\,\mathrm{dB} = -9.620399\,\mathrm{dB}.$$

Using the + sign in Eq. 7.7 and the − sign in Eq. 7.6, the first solution is (Fig. 7.2)

$$C_1 = 141.9609\,\mathrm{pF}, \qquad L_2 = 30.37537\,\mathrm{mH}.$$

Using the − sign in Eq. 7.7 and the + sign in Eq. 7.6, the second solution is (Fig. 7.3)

$$C_1 = 84.06313\,\mathrm{pF}, \qquad L_2 = 51.29617\,\mathrm{mH}.$$

Both of these circuit configurations satisfy the same design requirements exactly.

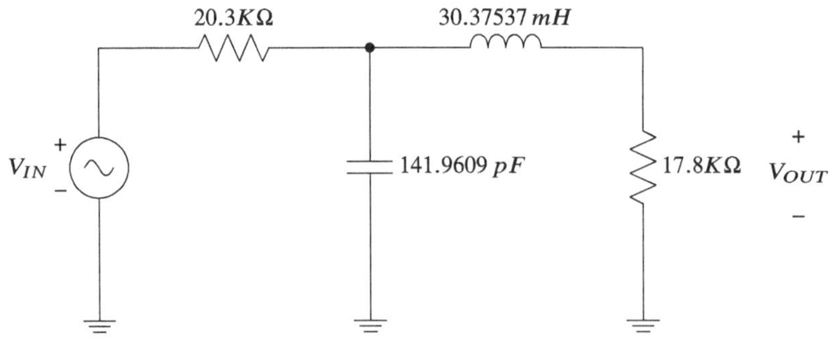

Fig. 7.2 First exact shunt-first low-pass solution

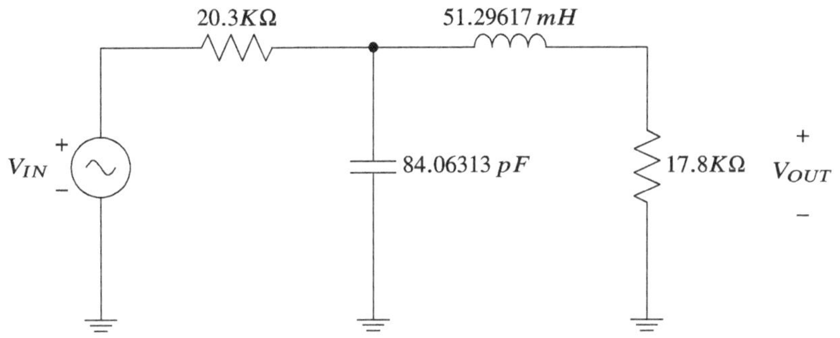

Fig. 7.3 Second exact shunt-first low-pass solution

7.2.2 Design Example—Second-Order Low-Pass Shunt-First Using MATLAB

A MATLAB implementation of the system of equations in Eq. 7.3 using the same design requirements may be written as

```
format longEng

% Shunt-first to be used for Rs > Rl
Rs = 20.3e3;
Rl = 17.8e3;
f  = 105e3;
w  = 2*pi*f;

syms C1 L2

E0 = (Rs/Rl)+1;
E1 = w*(C1*Rs + L2/Rl) ==
E0*(cos(0)/sin((1*pi)/4));
E2 = w^2*C1*L2*(Rs/Rl) ==
E0*((cos(0)/sin((1*pi)/4))*
(cos((1*pi)/4)/sin((2*pi)/4)));

result = solve(E1,E2);

C1 = vpa(result.C1)
L2 = vpa(result.L2)
```

The complete candidate solution tuple output for this design example is provided in Appendix B, Sect. B.1.

These match the quadratic results above, confirming that both real solutions are exact valid realizations.

7.3 Low-Pass Series-First

The transfer function for the second-order series-first doubly-terminated low-pass filter in Fig. 7.4 is

$$\frac{V_{OUT}}{V_{IN}} = \frac{1}{\omega^2 L_1 C_2 + \omega\left(\frac{L_1}{R_L} + C_2 R_S\right) + \left(\frac{R_S}{R_L} + 1\right)}. \tag{7.8}$$

The corresponding normalized Butterworth transfer function is

$$\frac{V_{OUT}}{V_{IN}} = \frac{1}{s^2\left(\left(\frac{\cos(0)}{\sin(\frac{1}{4}\pi)}\right)\left(\frac{\cos(\frac{1}{4}\pi)}{\sin(\frac{2}{4}\pi)}\right)\right) + s\left(\frac{\cos(0)}{\sin(\frac{1}{4}\pi)}\right) + 1}. \tag{7.9}$$

The second-degree term of the circuit denominator,

$$\omega^2 L_1 C_2,$$

is associated with the second-degree Butterworth term

$$s^2\left(\left(\frac{\cos(0)}{\sin(\frac{1}{4}\pi)}\right)\left(\frac{\cos(\frac{1}{4}\pi)}{\sin(\frac{2}{4}\pi)}\right)\right).$$

The first-degree term of the circuit denominator,

$$\omega\left(\frac{L_1}{R_L} + C_2 R_S\right),$$

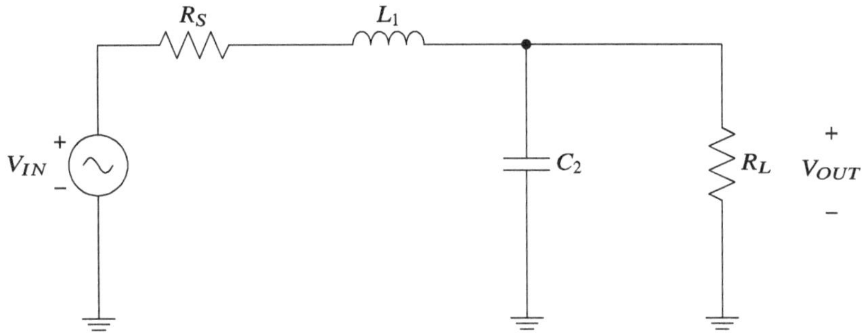

Fig. 7.4 Second-order series-first doubly-terminated low-pass filter for $R_S < R_L$

is associated with the first-degree Butterworth term

$$s\left(\frac{\cos(0)}{\sin(\frac{1}{4}\pi)}\right).$$

Applying the transformation rule gives

$$\begin{aligned} AC_{aspect} &= \left(DC_{aspect}\right)\left(Butterworth_{coef}\right) \\ \omega^2 L_1 C_2 &= \left(\frac{R_S}{R_L}+1\right)\left(\left(\frac{\cos(0)}{\sin(\frac{1}{4}\pi)}\right)\left(\frac{\cos(\frac{1}{4}\pi)}{\sin(\frac{2}{4}\pi)}\right)\right) \\ \omega\left(\frac{L_1}{R_L}+C_2 R_S\right) &= \left(\frac{R_S}{R_L}+1\right)\left(\frac{\cos(0)}{\sin(\frac{1}{4}\pi)}\right) \end{aligned} \tag{7.10}$$

This simplifies to

$$\omega^2 L_1 C_2 = \left(\frac{R_S}{R_L}+1\right), \tag{7.11}$$

and

$$\omega\left(\frac{L_1}{R_L}+C_2 R_S\right) = \left(\frac{R_S}{R_L}+1\right)(\sqrt{2}) = \left(\frac{R_S}{R_L}+1\right)\left(\csc\left(\frac{\pi}{4}\right)\right). \tag{7.12}$$

These resolve into the same independent expressions, Eqs. 7.6 and 7.7. Thus, the series-first second-order case also yields two candidate exact realizations, provided the corresponding solutions are real.

7.3.1 Design Example—Second-Order Low-Pass Series-First Using the Quadratic Forms

The design requirements are:

- $R_S = 46.37\,\text{k}\Omega$
- $R_L = 172.89\,\text{k}\Omega$
- $f_c = 372\,\text{kHz}$
- $\omega_c = 2.337344 \times 10^6\,\text{rad/s}$.

The gain at DC, or throughout the passband as established by the terminations, is

$$20\log_{10}\left(\frac{172.89\,\text{k}\Omega}{46.37\,\text{k}\Omega + 172.89\,\text{k}\Omega}\right) = -2.063791\,\text{dB}.$$

Therefore the gain at cutoff is

$$20\log_{10}\left(\frac{1}{\sqrt{2}} * \frac{172.89\,\text{k}\Omega}{172.89\,\text{k}\Omega + 46.37\,\text{k}\Omega}\right) = -5.074091\,\text{dB},$$

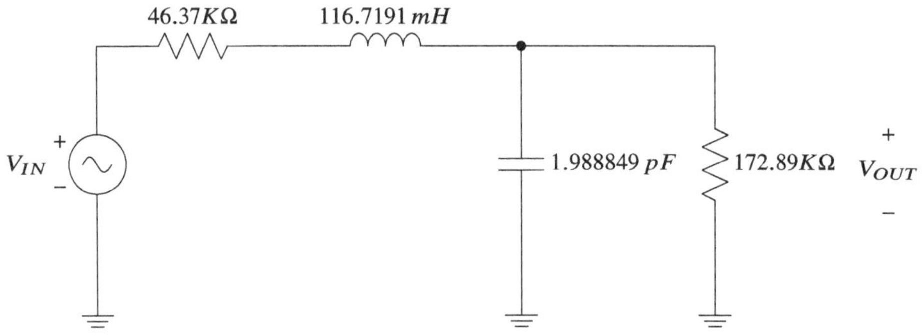

Fig. 7.5 First exact series-first low-pass solution

or

$$-2.063791\,\text{dB} - 3.010299\,\text{dB} = -5.074091\,\text{dB}.$$

Using the + sign in Eq. 7.6 and the − sign in Eq. 7.7, the first solution is (Fig. 7.5)

$$L_1 = 116.7191\,\text{mH}, \qquad C_2 = 1.988849\,\text{pF}.$$

Using the − sign in Eq. 7.6 and the + sign in Eq. 7.7, the second solution is (Fig. 7.6)

$$L_1 = 15.94442\,\text{mH}, \qquad C_2 = 14.55911\,\text{pF}.$$

Both circuit configurations satisfy the same Butterworth design requirements exactly.

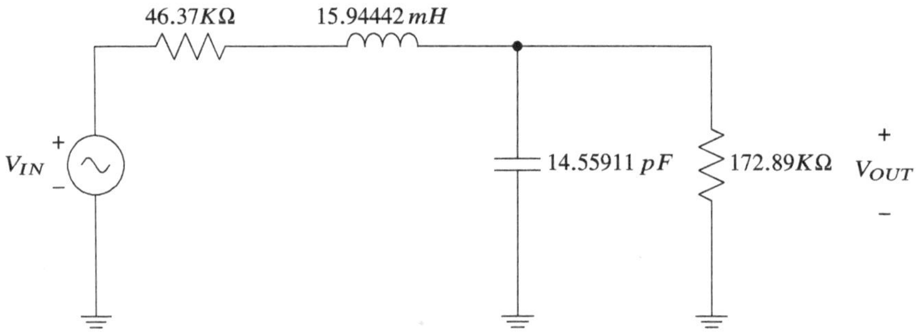

Fig. 7.6 Second exact series-first low-pass solution

7.3.2 Design Example—Second-Order Low-Pass Series-First Using MATLAB

A MATLAB implementation of the system of equations in Eq. 7.10 using the same design requirements may be written as

```
format longEng

% Series-first to be used for Rs < Rl
Rs = 46.37e3;
Rl = 172.89e3;
f  = 372e3;
w  = 2*pi*f;

syms L1 C2

E0 = 1/(Rl/(Rs+Rl));
E1 = w*(L1/Rl + C2*Rs) ==
E0*(cos(0)/sin((1*pi)/4));
E2 = w^2*L1*C2 == E0;

result = solve(E1,E2);

L1 = vpa(result.L1)
C2 = vpa(result.C2)
```

The complete candidate solution tuple output for this design example is provided in Appendix B, Sect. B.2.

These match the quadratic results above, again confirming the existence of two exact real realizations.

7.4 Time Constants

Equations 6.8 and 6.17 define τ_C and τ_L, which are used throughout this section.

For the shunt-first second-order low-pass filter, the first-degree portion of the system is

$$\begin{aligned} \omega\left(C_1 R_S + \frac{L_2}{R_L}\right) &= \left(\frac{R_S}{R_L} + 1\right)(\sqrt{2}) \\ \omega\left(C_1 \frac{R_S R_L}{R_S + R_L} + \frac{L_2}{R_S + R_L}\right) &= \sqrt{2}. \end{aligned} \tag{7.13}$$

For the series-first second-order low-pass filter, the first-degree portion is

$$\begin{aligned} \omega\left(\frac{L_1}{R_L} + C_2 R_S\right) &= \left(\frac{R_S}{R_L} + 1\right)(\sqrt{2}) \\ \omega\left(\frac{L_1}{R_S + R_L} + C_2 \frac{R_S R_L}{R_S + R_L}\right) &= \sqrt{2}. \end{aligned} \tag{7.14}$$

Thus, both second-order topologies resolve to

$$\omega\left(\tau_L + \tau_C\right) = \sqrt{2} = \frac{1}{\sin(\frac{1}{4}\pi)},$$

or equivalently,

$$\omega\tau_L + \omega\tau_C = \frac{1}{\sin(\frac{1}{4}\pi)}. \tag{7.15}$$

If $R_S = R_L = 1$, then

$$\omega\tau_L = \omega\tau_C = \sin\left(\frac{1}{4}\pi\right) = \sin\left(\frac{3}{4}\pi\right),$$

and therefore

$$\sin\left(\frac{1}{4}\pi\right) + \sin\left(\frac{3}{4}\pi\right) = \frac{1}{\sin(\frac{1}{4}\pi)} = \csc\left(\frac{1}{4}\pi\right). \tag{7.16}$$

If $R_S \neq R_L$, then

$$\omega\tau_L \neq \omega\tau_C \neq \sin\left(\frac{1}{4}\pi\right), \tag{7.17}$$

yet the summed relationship of Eq. 7.15 remains valid.

This is the first higher-order instance of the finite trigonometric time-constant pattern that is later collected in Chap. 12.

7.5 High-Pass Shunt-First

A second-order shunt-first high-pass filter is obtained by first synthesizing the corresponding second-order shunt-first low-pass ladder and then converting each reactive element using Eqs. 6.19 and 6.20. Therefore, the high-pass example is not a separate coefficient-matching derivation. It is a transformation example that uses the shunt-first low-pass system of Eq. 7.3 to obtain a low-pass precursor, which is then converted into the final high-pass ladder. The low-pass shunt capacitor becomes a high-pass shunt inductor, and the low-pass series inductor becomes a high-pass series capacitor.

7.5.1 Design Example—Second-Order High-Pass Shunt-First

The design requirements are:

- $R_S = 142.3\,\text{k}\Omega$
- $R_L = 72.9\,\text{k}\Omega$
- $f_c = 37.57\,\text{kHz}$
- $\omega_c = 236.0592 \times 10^3\,\text{rad/s}$.

The passband gain established by the terminations is

$$20\log_{10}\left(\frac{72.9\,\text{k}\Omega}{142.3\,\text{k}\Omega + 72.9\,\text{k}\Omega}\right) = -9.402295\,\text{dB}.$$

Therefore the gain at the high-pass cut-on frequency is

$$20\log_{10}\left(\frac{1}{\sqrt{2}} * \frac{72.9\,\text{k}\Omega}{72.9\,\text{k}\Omega + 142.3\,\text{k}\Omega}\right) = -12.412595\,\text{dB},$$

or

$$-9.402295\,\text{dB} - 3.010299\,\text{dB} = -12.412595\,\text{dB}.$$

To obtain the low-pass precursor for the high-pass realization, the shunt-first low-pass system of Eq. 7.3 is solved using the high-pass design requirements. A MATLAB implementation may be written as

```
format longEng

% Shunt-first to be used for Rs > Rl
Rs = 142.3e3;
Rl = 72.9e3;
f  = 37.57e3;
w  = 2*pi*f;

syms C1 L2

E0 = (Rs/Rl)+1;
E1 = w*(C1*Rs + L2/Rl) ==
E0*(cos(0)/sin((1*pi)/4));
E2 = w^2*C1*L2*(Rs/Rl) ==
E0*((cos(0)/sin((1*pi)/4))*
(cos((1*pi)/4)/sin((2*pi)/4)));

result = solve(E1,E2);

C1 = vpa(result.C1)
L2 = vpa(result.L2)
```

The complete candidate solution tuple output for the low-pass precursor solve used in this high-pass design example is provided in Appendix B, Sect. B.3.

In this example, the solver returns two fully real candidate tuples. Either real solution satisfies the design target exactly. Using the first solution gives (Fig. 7.7)

$$C_1 = 97.428795\,\text{pF}, \qquad L_2 = 278.55304\,\text{mH}.$$

The high-pass shunt inductor becomes

$$184.1919\,\text{mH} = \frac{1}{(236.0592 \times 10^3)^2(97.428795\,\text{pF})},$$

and the high-pass series capacitor becomes (Fig. 7.8)

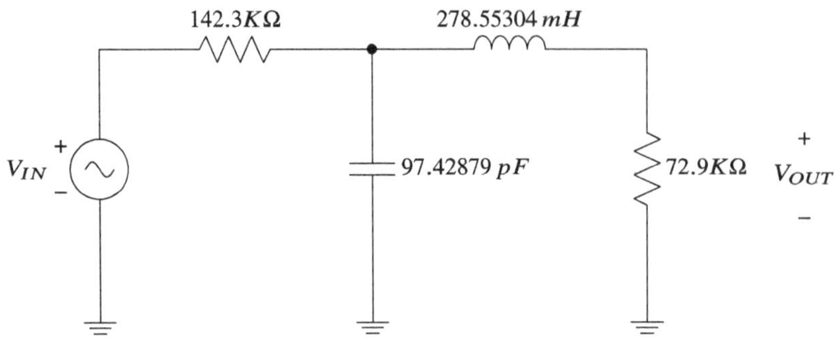

Fig. 7.7 Second-order shunt-first low-pass precursor for high-pass transformation

$$64.42433\,\text{pF} = \frac{1}{(236.0592 \times 10^3)^2(278.55304\,\text{mH})}.$$

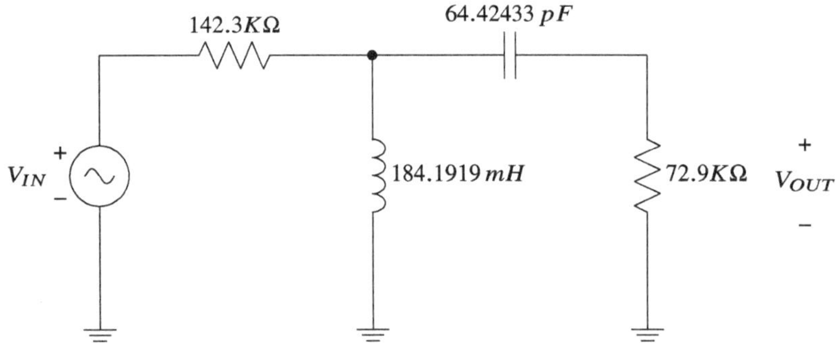

Fig. 7.8 Second-order shunt-first high-pass solution

7.6 High-Pass Series-First

A second-order series-first high-pass filter is obtained by first synthesizing the corresponding second-order series-first low-pass ladder and then converting each reactive element using Eqs. 6.19 and 6.20. Therefore, the high-pass example is not a separate coefficient-matching derivation. It is a transformation example that uses the series-first low-pass system of Eq. 7.10 to obtain a low-pass precursor, which is then converted into the final high-pass ladder. The low-pass series inductor becomes a high-pass series capacitor, and the low-pass shunt capacitor becomes a high-pass shunt inductor.

7.6.1 Design Example—Second-Order High-Pass Series-First

The design requirements are:

- $R_S = 17\,\mathrm{k\Omega}$
- $R_L = 17.5\,\mathrm{k\Omega}$
- $f_c = 1.25\,\mathrm{kHz}$
- $\omega_c = 7.853981 \times 10^3\,\mathrm{rad/s}$.

The passband gain established by the terminations is

$$20\log_{10}\left(\frac{17.5\,\mathrm{k\Omega}}{17\,\mathrm{k\Omega} + 17.5\,\mathrm{k\Omega}}\right) = -5.895621\,\mathrm{dB}.$$

Therefore the gain at the high-pass cut-on frequency is

$$20\log_{10}\left(\frac{1}{\sqrt{2}} * \frac{17.5\,\mathrm{k\Omega}}{17.5\,\mathrm{k\Omega} + 17\,\mathrm{k\Omega}}\right) = -8.905921\,\mathrm{dB},$$

or

$$-5.895621\,\mathrm{dB} - 3.010299\,\mathrm{dB} = -8.905921\,\mathrm{dB}.$$

To obtain the low-pass precursor for the high-pass realization, the series-first low-pass system of Eq. 7.10 is solved using the high-pass design requirements. A MATLAB implementation may be written as

```
format longEng

% Series-first to be used for Rs < Rl
Rs = 17e3;
Rl = 17.5e3;
f  = 1.25e3;
w  = 2*pi*f;

syms L1 C2

E0 = 1/(Rl/(Rs+Rl));
E1 = w*(L1/Rl + C2*Rs) ==
E0*(cos(0)/sin((1*pi)/4));
E2 = w^2*L1*C2 == E0;

result = solve(E1,E2);

L1 = vpa(result.L1)
C2 = vpa(result.C2)
```

The complete candidate solution tuple output for the low-pass precursor solve used in this high-pass design example is provided in Appendix B, Sect. B.4.

In this example, the solver returns two fully real candidate tuples. Using the second real low-pass solution gives (Fig. 7.9)

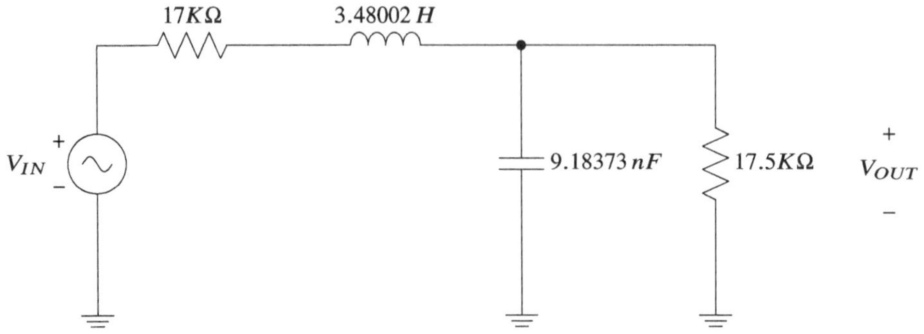

Fig. 7.9 Second-order series-first low-pass precursor for high-pass transformation

$$L_1 = 3.4800207\,\text{H}, \qquad C_2 = 9.18373728\,\text{nF}.$$

The corresponding high-pass elements are

$$4.658417\,\text{nF} = \frac{1}{(7.853981 \times 10^3)^2(3.4800207\,\text{H})},$$

for the series capacitor, and

$$1.765227\,\text{H} = \frac{1}{(7.853981 \times 10^3)^2(9.18373728\,\text{nF})},$$

for the shunt inductor (Fig. 7.10).

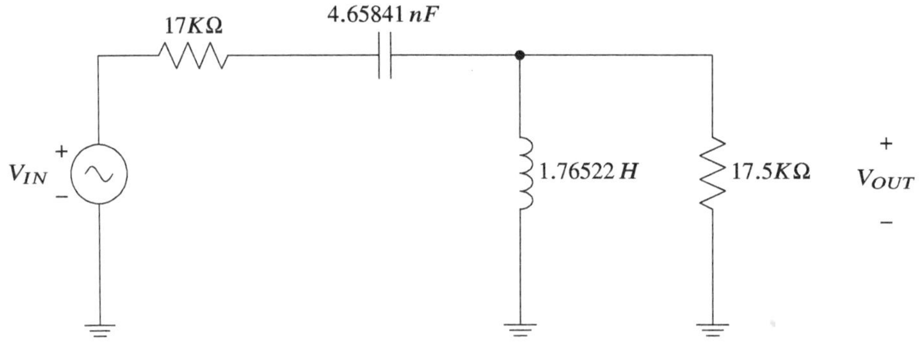

Fig. 7.10 Second-order series-first high-pass solution

7.7 Conclusion

The second-order case is the first order at which the synthesis method yields more than one exact real passive realization for the same Butterworth design target. This makes the chapter especially important within the logic of the monograph: it confirms that exact synthesis need not imply uniqueness of realization.

It also shows how topology selection, solution pairing, and time-constant structure begin to interact once the filter order exceeds one. These same themes recur in the higher-order chapters, but in richer form.

Accordingly, Chap. 8 turns to the third-order case, where odd-order symmetry appears in a more developed denominator structure and the solution set expands to include both multiple real realizations and non-real candidate tuples.

3rd-Order Doubly-Terminated Passive Butterworth Filter Network Synthesis

8

Odd order begins to show the full richness of symmetry.

Chapter 7 showed that exact synthesis need not imply a unique passive realization. At second order, multiple exact real solutions appear for the first time, and topology selection becomes constrained by even-order asymmetry. The present chapter extends the same synthesis framework to third order, where the odd-order symmetry discussed earlier becomes visible in a richer denominator structure and the candidate solution set expands further.

As in the preceding chapters, the method remains unchanged in principle: each degree-k frequency-dependent term of the realized ladder denominator is set equal, at cutoff, to the frequency-independent term multiplied by the corresponding normalized Butterworth coefficient. What changes at third order is the structure of the denominator and the richness of the resulting solution tuples.

8.1 Odd Symmetry

Third-order doubly-terminated passive Butterworth ladder filters are odd-order and structurally symmetric. Accordingly, either a shunt-first topology or a series-first topology may be used regardless of the relative values of R_S and R_L.

In the third-order case, the mapped system contains three unknown reactive element values and may produce multiple candidate solution tuples. In the examples that follow, the valid passive realizations are the real-valued solutions; complex-valued solution tuples are not retained as physical filter realizations in the present work.

K. M. P. Ryan, *Polynomial Symmetry in Butterworth Filter Synthesis*,
Synthesis Lectures on Engineering, Science, and Technology,
https://doi.org/10.1007/978-3-032-33468-8_8

8.2 Low-Pass Shunt-First

The transfer function for the third-order shunt-first doubly-terminated low-pass Butterworth filter in Fig. 8.1 is

$$\frac{V_{OUT}}{V_{IN}} = \frac{1}{\begin{array}{l}\omega^3 C_1 L_2 C_3 R_S \\ + \omega^2 \left(C_1 L_2 \frac{R_S}{R_L} + L_2 C_3\right) \\ + \omega \left(C_1 R_S + \frac{L_2}{R_L} + C_3 R_S\right) + \left(\frac{R_S}{R_L} + 1\right)\end{array}} \tag{8.1}$$

The corresponding normalized Butterworth transfer function using the product formula of Eq. 2.1 is

$$\frac{V_{OUT}}{V_{IN}} = \frac{1}{\begin{array}{l} s^3 \left(\left(\frac{\cos(0)}{\sin(\frac{1}{6}\pi)}\right)\left(\frac{\cos(\frac{1}{6}\pi)}{\sin(\frac{2}{6}\pi)}\right)\left(\frac{\cos(\frac{2}{6}\pi)}{\sin(\frac{3}{6}\pi)}\right)\right) \\ + s^2 \left(\left(\frac{\cos(0)}{\sin(\frac{1}{6}\pi)}\right)\left(\frac{\cos(\frac{1}{6}\pi)}{\sin(\frac{2}{6}\pi)}\right)\right) \\ + s \left(\frac{\cos(0)}{\sin(\frac{1}{6}\pi)}\right) + 1 \end{array}} \tag{8.2}$$

The third-degree term of the realized denominator,

$$\omega^3 C_1 L_2 C_3 R_S,$$

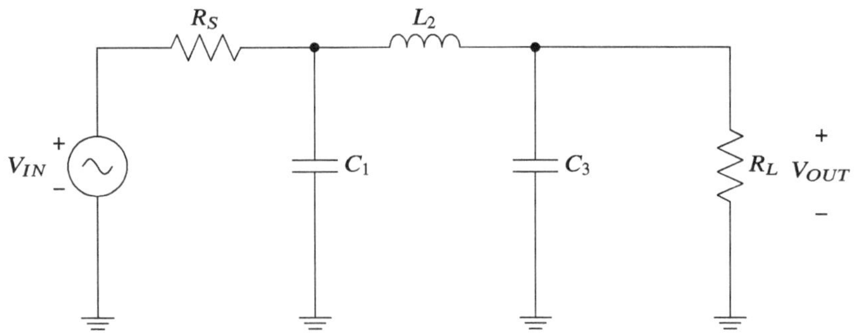

Fig. 8.1 Third-order shunt-first doubly-terminated low-pass Butterworth filter

is associated with the third-degree Butterworth term

$$s^3\left(\left(\frac{\cos(0)}{\sin(\frac{1}{6}\pi)}\right)\left(\frac{\cos(\frac{1}{6}\pi)}{\sin(\frac{2}{6}\pi)}\right)\left(\frac{\cos(\frac{2}{6}\pi)}{\sin(\frac{3}{6}\pi)}\right)\right).$$

The second-degree term of the realized denominator,

$$\omega^2\left(C_1L_2\frac{R_S}{R_L}+L_2C_3\right),$$

is associated with the second-degree Butterworth term

$$s^2\left(\left(\frac{\cos(0)}{\sin(\frac{1}{6}\pi)}\right)\left(\frac{\cos(\frac{1}{6}\pi)}{\sin(\frac{2}{6}\pi)}\right)\right).$$

The first-degree term of the realized denominator,

$$\omega\left(C_1R_S+\frac{L_2}{R_L}+C_3R_S\right),$$

is associated with the first-degree Butterworth term

$$s\left(\frac{\cos(0)}{\sin(\frac{1}{6}\pi)}\right).$$

Applying the transformation rule yields

$$\begin{aligned}
AC_{aspect} &= \left(DC_{aspect}\right)\left(Butterworth_{coef}\right)\\
\omega^3C_1L_2C_3R_S &= \left(\frac{R_S}{R_L}+1\right)\left(\left(\frac{\cos(0)}{\sin(\frac{1}{6}\pi)}\right)\left(\frac{\cos(\frac{1}{6}\pi)}{\sin(\frac{2}{6}\pi)}\right)\left(\frac{\cos(\frac{2}{6}\pi)}{\sin(\frac{3}{6}\pi)}\right)\right)\\
\omega^2\left(C_1L_2\frac{R_S}{R_L}+L_2C_3\right) &= \left(\frac{R_S}{R_L}+1\right)\left(\left(\frac{\cos(0)}{\sin(\frac{1}{6}\pi)}\right)\left(\frac{\cos(\frac{1}{6}\pi)}{\sin(\frac{2}{6}\pi)}\right)\right)\\
\omega\left(C_1R_S+\frac{L_2}{R_L}+C_3R_S\right) &= \left(\frac{R_S}{R_L}+1\right)\left(\frac{\cos(0)}{\sin(\frac{1}{6}\pi)}\right)
\end{aligned} \tag{8.3}$$

The system of Eq. 8.3 simplifies to

$$\omega^3C_1L_2C_3R_S=\left(\frac{R_S}{R_L}+1\right), \tag{8.4}$$

$$\omega^2\left(C_1L_2\frac{R_S}{R_L}+L_2C_3\right)=2\left(\frac{R_S}{R_L}+1\right), \tag{8.5}$$

and

$$\omega\left(C_1 R_S + \frac{L_2}{R_L} + C_3 R_S\right) = 2\left(\frac{R_S}{R_L} + 1\right). \tag{8.6}$$

The third-degree term and the constant term are equal, while the second- and first-degree terms are equal. This is the denominator-level manifestation of the odd-order symmetry discussed earlier in Sect. 2.3.2.1.

8.2.1 Design Example—Third-Order Low-Pass Shunt-First

The design requirements are:

- $R_S = 510\,\text{k}\Omega$
- $R_L = 511\,\text{k}\Omega$
- $f_c = 343\,\text{kHz}$
- $\omega_c = 2.155132 \times 10^6\,\text{rad/s}$

The gain at DC, or throughout the passband as established by the terminations, is

$$20\log_{10}\left(\frac{511\,\text{k}\Omega}{510\,\text{k}\Omega + 511\,\text{k}\Omega}\right) = -6.012096\,\text{dB}.$$

Therefore the gain at cutoff is

$$20\log_{10}\left(\frac{1}{\sqrt{2}} * \frac{511\,\text{k}\Omega}{511\,\text{k}\Omega + 510\,\text{k}\Omega}\right) = -9.022395\,\text{dB},$$

or

$$-6.012096\,\text{dB} - 3.010299\,\text{dB} = -9.022395\,\text{dB}.$$

A MATLAB implementation of the system of equations in Eq. 8.3 may be written as

```
format longEng

Rs = 510e3
Rl = 511e3
f = 343e3
w = 2*pi*f

syms C1 L2 C3

E0 = (Rs/Rl)+1
E1 = (w^1)*(C1*Rs+L2/Rl+C3*Rs) ==
E0*(cos(0)/sin((1*pi)/6))
E2 = (w^2)*(C1*L2*(Rs/Rl)+L2*C3) ==
E0*(cos(0)/sin((1*pi)/6))
E3 = (w^3)*C1*L2*C3*Rs == E0

result = solve(E1,E2,E3)

C1 = vpa(result.C1)
```

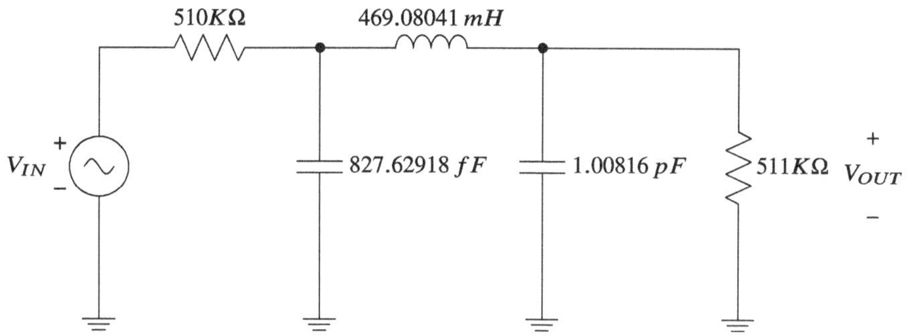

Fig. 8.2 First valid shunt-first low-pass solution

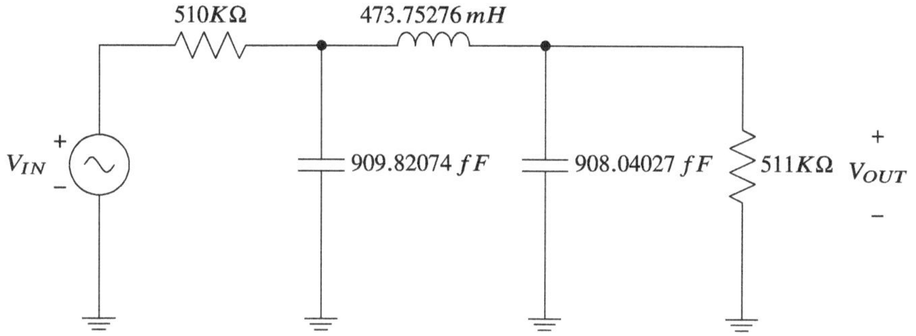

Fig. 8.3 Second valid shunt-first low-pass solution

```
L2 = vpa(result.L2)
C3 = vpa(result.C3)
```

The complete candidate solution tuple output for this design example is provided in Appendix B, Sect. B.5.

In this example, the solver returns four candidate tuples: two fully real tuples and one complex-conjugate pair. The two real tuples are retained as valid passive realizations.

The first valid solution is (Fig. 8.2)

$$C_1 = 827.6291871\,\text{fF}, \quad L_2 = 469.0804189\,\text{mH}, \quad C_3 = 1.00816031\,\text{pF}.$$

The second valid solution is (Fig. 8.3)

$$C_1 = 909.8207459\,\text{fF}, \quad L_2 = 473.75276\,\text{mH}, \quad C_3 = 908.0402748\,\text{fF}.$$

8.3 Low-Pass Series-First

The transfer function for the third-order series-first doubly-terminated low-pass Butterworth filter in Fig. 8.4 is

$$\frac{V_{OUT}}{V_{IN}} = \frac{1}{\begin{array}{l}\omega^3 L_1 C_2 L_3 \frac{1}{R_L} \\ + \omega^2 \left(L_1 C_2 + C_2 L_3 \frac{R_S}{R_L} \right) \\ + \omega \left(\frac{L_1}{R_L} + C_2 R_S + \frac{L_3}{R_L} \right) + \left(\frac{R_S}{R_L} + 1 \right)\end{array}} \tag{8.7}$$

The corresponding normalized Butterworth transfer function is

$$\frac{V_{OUT}}{V_{IN}} = \frac{1}{s^3 \left(\left(\frac{\cos(0)}{\sin(\frac{1}{6}\pi)} \right) \left(\frac{\cos(\frac{1}{6}\pi)}{\sin(\frac{2}{6}\pi)} \right) \left(\frac{\cos(\frac{2}{6}\pi)}{\sin(\frac{3}{6}\pi)} \right) \right) + s^2 \left(\left(\frac{\cos(0)}{\sin(\frac{1}{6}\pi)} \right) \left(\frac{\cos(\frac{1}{6}\pi)}{\sin(\frac{2}{6}\pi)} \right) \right) + s \left(\frac{\cos(0)}{\sin(\frac{1}{6}\pi)} \right) + 1} \tag{8.8}$$

The third-degree term of the realized denominator,

$$\omega^3 L_1 C_2 L_3 \frac{1}{R_L},$$

is associated with the third-degree Butterworth term

$$s^3 \left(\left(\frac{\cos(0)}{\sin(\frac{1}{6}\pi)} \right) \left(\frac{\cos(\frac{1}{6}\pi)}{\sin(\frac{2}{6}\pi)} \right) \left(\frac{\cos(\frac{2}{6}\pi)}{\sin(\frac{3}{6}\pi)} \right) \right).$$

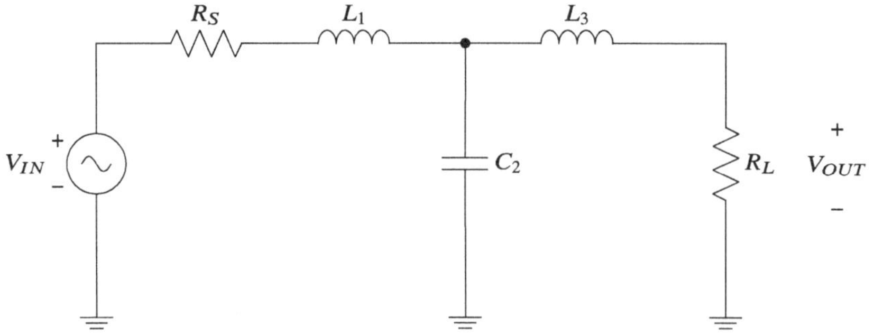

Fig. 8.4 Third-order series-first doubly-terminated low-pass Butterworth filter

The second-degree term of the realized denominator,

$$\omega^2\left(L_1C_2 + C_2L_3\frac{R_S}{R_L}\right),$$

is associated with the second-degree Butterworth term

$$s^2\left(\left(\frac{\cos(0)}{\sin(\frac{1}{6}\pi)}\right)\left(\frac{\cos(\frac{1}{6}\pi)}{\sin(\frac{2}{6}\pi)}\right)\right).$$

The first-degree term of the realized denominator,

$$\omega\left(\frac{L_1}{R_L} + C_2R_S + \frac{L_3}{R_L}\right),$$

is associated with the first-degree Butterworth term

$$s\left(\frac{\cos(0)}{\sin(\frac{1}{6}\pi)}\right).$$

Applying the transformation rule yields

$$\begin{aligned}
AC_{aspect} &= \left(DC_{aspect}\right)\left(Butterworth_{coef}\right)\\
\omega^3L_1C_2L_3\frac{1}{R_L} &= \left(\frac{R_S}{R_L}+1\right)\left(\left(\frac{\cos(0)}{\sin(\frac{1}{6}\pi)}\right)\left(\frac{\cos(\frac{1}{6}\pi)}{\sin(\frac{2}{6}\pi)}\right)\left(\frac{\cos(\frac{2}{6}\pi)}{\sin(\frac{3}{6}\pi)}\right)\right)\\
\omega^2\left(L_1C_2 + C_2L_3\frac{R_S}{R_L}\right) &= \left(\frac{R_S}{R_L}+1\right)\left(\left(\frac{\cos(0)}{\sin(\frac{1}{6}\pi)}\right)\left(\frac{\cos(\frac{1}{6}\pi)}{\sin(\frac{2}{6}\pi)}\right)\right)\\
\omega\left(\frac{L_1}{R_L} + C_2R_S + \frac{L_3}{R_L}\right) &= \left(\frac{R_S}{R_L}+1\right)\left(\frac{\cos(0)}{\sin(\frac{1}{6}\pi)}\right)
\end{aligned} \tag{8.9}$$

The system of Eq. 8.9 simplifies to

$$\omega^3L_1C_2L_3\frac{1}{R_L} = \left(\frac{R_S}{R_L}+1\right), \tag{8.10}$$

$$\omega^2\left(L_1C_2 + C_2L_3\frac{R_S}{R_L}\right) = 2\left(\frac{R_S}{R_L}+1\right), \tag{8.11}$$

and

$$\omega\left(\frac{L_1}{R_L} + C_2R_S + \frac{L_3}{R_L}\right) = 2\left(\frac{R_S}{R_L}+1\right). \tag{8.12}$$

As in the shunt-first case, the third-degree term and the constant term are equal, while the second- and first-degree terms are equal.

8.3.1 Design Example—Third-Order Low-Pass Series-First

The design requirements are:

- $R_S = 10\,\mathrm{k}\Omega$
- $R_L = 8.75\,\mathrm{k}\Omega$
- $f_c = 43.7\,\mathrm{kHz}$
- $\omega_c = 274.5751 \times 10^3\,\mathrm{rad/s}$

The gain at DC, or throughout the passband as established by the terminations, is

$$20\log_{10}\left(\frac{8.75\,\mathrm{k}\Omega}{10\,\mathrm{k}\Omega + 8.75\,\mathrm{k}\Omega}\right) = -6.619864\,\mathrm{dB}.$$

Therefore the gain at cutoff is

$$20\log_{10}\left(\frac{1}{\sqrt{2}} * \frac{8.75\,\mathrm{k}\Omega}{8.75\,\mathrm{k}\Omega + 10\,\mathrm{k}\Omega}\right) = -9.630163\,\mathrm{dB},$$

or

$$-6.619864\,\mathrm{dB} - 3.010299\,\mathrm{dB} = -9.630163\,\mathrm{dB}.$$

A MATLAB implementation of the system of equations in Eq. 8.9 may be written as

```
format longEng

Rs = 10e3;
Rl = 8.75e3;
f  = 43.7e3;
w  = 2*pi*f;

syms L1 C2 L3

E0 = (w^0)*((Rs/Rl)+1)
E1 = (w^1)*(L1/Rl + C2*Rs + L3/Rl) ==
E0 * (cos(0)/sin((1*pi)/6))
E2 = (w^2)*(L1*C2 + C2*L3*(Rs/Rl)) ==
E0 * (cos(0)/sin((1*pi)/6))
E3 = (w^3)*(L1*C2*L3)/Rl == E0

result = solve(E1,E2,E3)

L1 = vpa(result.L1)
C2 = vpa(result.C2)
L3 = vpa(result.L3)
```

The complete candidate solution tuple output for this design example is provided in Appendix B, Sect. B.6.

The first two candidate tuples form a complex-conjugate pair and are not retained as passive realizations. The last two tuples are fully real and therefore valid.

The first valid real solution is (Fig. 8.5)

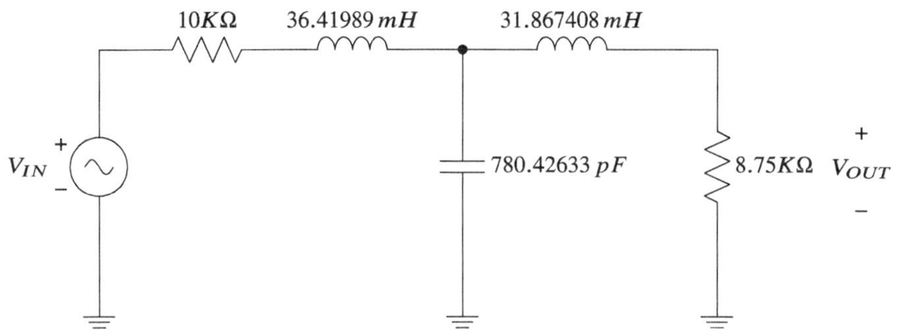

Fig. 8.5 First valid series-first low-pass solution

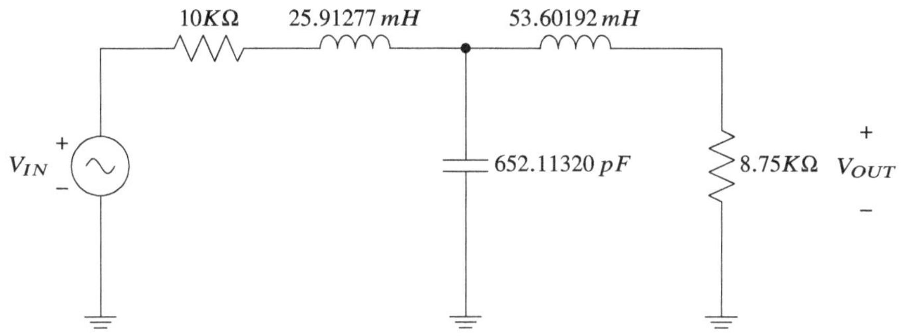

Fig. 8.6 Second valid series-first low-pass solution

$$L_1 = 36.419895\,\text{mH}, \qquad C_2 = 780.42633095\,\text{pF}, \qquad L_3 = 31.8674085138\,\text{mH}.$$

The second valid real solution is (Fig. 8.6)

$$L_1 = 25.91277855\,\text{mH}, \qquad C_2 = 652.113208\,\text{pF}, \qquad L_3 = 53.601923644\,\text{mH}.$$

8.4 Time Constants

Equations 6.8 and 6.17 define τ_C and τ_L, which are used throughout this section.

For the shunt-first third-order low-pass filter, the first-degree portion of the system is

$$\begin{aligned} \omega\left(C_1 R_S + \frac{L_2}{R_L} + C_3 R_S\right) &= \left(\frac{R_S}{R_L} + 1\right)(2) \\ \omega\left(C_1 \frac{R_S R_L}{R_S + R_L} + \frac{L_2}{R_S + R_L} + C_3 \frac{R_S R_L}{R_S + R_L}\right) &= 2. \end{aligned} \tag{8.13}$$

For the series-first third-order low-pass filter, the first-degree portion is

$$\omega\left(\frac{L_1}{R_L} + C_2 R_S + \frac{L_3}{R_L}\right) = \left(\frac{R_S}{R_L} + 1\right)(2)$$
$$\omega\left(\frac{L_1}{R_S + R_L} + C_2\frac{R_S R_L}{R_S + R_L} + \frac{L_3}{R_S + R_L}\right) = 2. \tag{8.14}$$

Both topologies therefore resolve to

$$\omega\left(\tau_{C_1} + \tau_{L_2} + \tau_{C_3}\right) = \omega\left(\tau_{L_1} + \tau_{C_2} + \tau_{L_3}\right) = 2 = \frac{1}{\sin(\frac{1}{6}\pi)}.$$

Equivalently,

$$\omega\tau_{C_1} + \omega\tau_{L_2} + \omega\tau_{C_3} = \omega\tau_{L_1} + \omega\tau_{C_2} + \omega\tau_{L_3} = \frac{1}{\sin(\frac{1}{6}\pi)}. \tag{8.15}$$

If $R_S = R_L = 1$, then

$$\omega\tau_{C_1} = \omega\tau_{L_1} = \sin\left(\frac{1}{6}\pi\right),$$
$$\omega\tau_{L_2} = \omega\tau_{C_2} = \sin\left(\frac{3}{6}\pi\right),$$
$$\omega\tau_{C_3} = \omega\tau_{L_3} = \sin\left(\frac{5}{6}\pi\right),$$

so that

$$\sin\left(\frac{1}{6}\pi\right) + \sin\left(\frac{3}{6}\pi\right) + \sin\left(\frac{5}{6}\pi\right) = \frac{1}{\sin(\frac{1}{6}\pi)} = \csc\left(\frac{1}{6}\pi\right). \tag{8.16}$$

If $R_S \neq R_L$, then

$$\omega\tau_{C_1} \neq \omega\tau_{L_1} \neq \sin\left(\frac{1}{6}\pi\right),$$
$$\omega\tau_{L_2} \neq \omega\tau_{C_2} \neq \sin\left(\frac{3}{6}\pi\right), \tag{8.17}$$
$$\omega\tau_{C_3} \neq \omega\tau_{L_3} \neq \sin\left(\frac{5}{6}\pi\right),$$

yet the summed identity of Eq. 8.16 remains valid.

This is the next higher-order instance of the finite trigonometric time-constant structure that is later collected in Chap. 12.

8.5 High-Pass Shunt-First

A third-order shunt-first high-pass filter is obtained by first synthesizing the corresponding third-order shunt-first low-pass ladder and then converting each reactive element using Eqs. 6.19 and 6.20. Therefore, the high-pass example is not a separate coefficient-matching derivation. It is a transformation example that uses the shunt-first low-pass system of Eq. 8.3 to obtain a low-pass precursor, which is then converted into the final high-pass ladder. Low-pass shunt capacitors become high-pass shunt inductors, and the low-pass series inductor becomes a high-pass series capacitor.

8.5.1 Design Example—Third-Order High-Pass Shunt-First

The design requirements are:

- $R_S = 14.83\,\text{k}\Omega$
- $R_L = 176.47\,\text{k}\Omega$
- $f_c = 145.3\,\text{kHz}$
- $\omega_c = 912.9468 \times 10^3\,\text{rad/s}$

The passband gain established by the terminations is

$$20\log_{10}\left(\frac{176.47\,\text{k}\Omega}{14.83\,\text{k}\Omega + 176.47\,\text{k}\Omega}\right) = -0.700882\,\text{dB}.$$

Therefore the gain at the high-pass cut-on frequency is

$$20\log_{10}\left(\frac{1}{\sqrt{2}} * \frac{176.47\,\text{k}\Omega}{176.47\,\text{k}\Omega + 14.83\,\text{k}\Omega}\right) = -3.711182\,\text{dB},$$

or

$$-0.700882\,\text{dB} - 3.010299\,\text{dB} = -3.711182\,\text{dB}.$$

To obtain the low-pass precursor for the high-pass realization, the shunt-first low-pass system of Eq. 8.3 is solved using the high-pass design requirements. A MATLAB implementation may be written as

```
format longEng

Rs = 14.83e3
Rl = 176.47e3
f = 145.3e3
w = 2*pi*f

syms C1 L2 C3

E0 = (Rs/Rl)+1
```

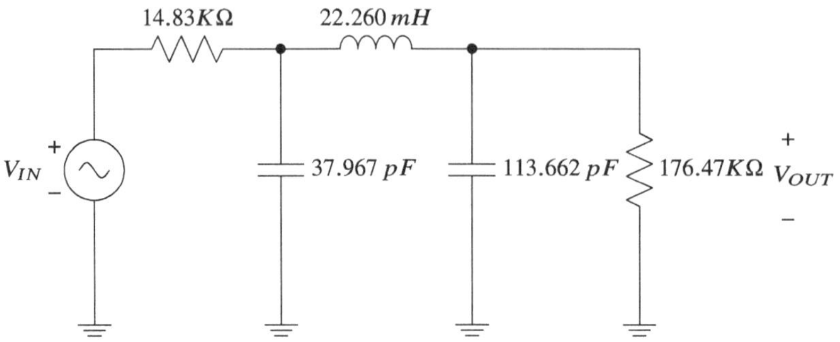

Fig. 8.7 Representative shunt-first low-pass precursor

```
E1 = (w^1)*(C1*Rs+L2/R1+C3*Rs) ==
E0*(cos(0)/sin((1*pi)/6))
E2 = (w^2)*(C1*L2*(Rs/R1)+L2*C3) ==
E0*(cos(0)/sin((1*pi)/6))
E3 = (w^3)*C1*L2*C3*Rs == E0

result = solve(E1,E2,E3)

C1 = vpa(result.C1)
L2 = vpa(result.L2)
C3 = vpa(result.C3)
```

The complete candidate solution tuple output for the low-pass precursor solve used in this high-pass design example is provided in Appendix B, Sect. B.7.

For this low-pass precursor solve, the solver returns four candidate tuples: two fully real tuples and one complex-conjugate pair. The two fully real tuples are retained as valid passive realizations.

To provide a concrete component set for the high-pass transformation, one of the fully real tuples is selected as a representative low-pass precursor. The first fully real tuple is used here for the transformation and verification steps (Fig. 8.7):

$$C_1 = 37.96702323\,\text{pF}, \qquad L_2 = 22.26094128\,\text{mH}, \qquad C_3 = 113.66229\,\text{pF}.$$

The corresponding high-pass elements are

$$31.60112\,\text{mH} = \frac{1}{(912.9468 \times 10^3)^2(37.96702323\,\text{pF})},$$

for the first shunt inductor,

$$53.89711\,\text{pF} = \frac{1}{(912.9468 \times 10^3)^2(22.26094128\,\text{mH})},$$

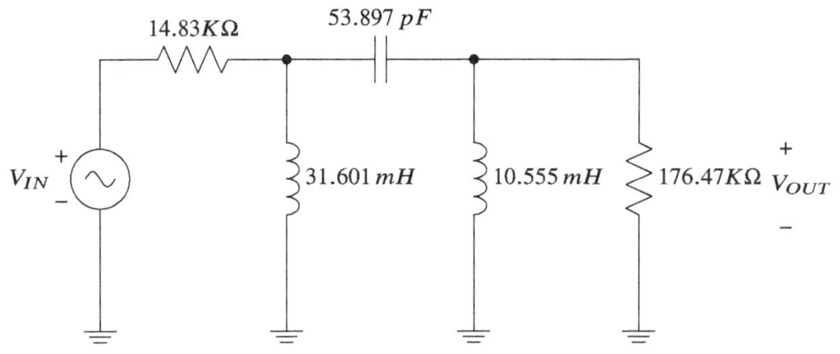

Fig. 8.8 Representative shunt-first high-pass solution

for the series capacitor, and

$$10.55583\,\text{mH} = \frac{1}{(912.9468 \times 10^3)^2(113.66229\,\text{pF})},$$

for the second shunt inductor (Fig. 8.8).

8.6 High-Pass Series-First

A third-order series-first high-pass filter is obtained by first synthesizing the corresponding third-order series-first low-pass ladder and then converting each reactive element using Eqs. 6.19 and 6.20. Therefore, the high-pass example is not a separate coefficient-matching derivation. It is a transformation example that uses the series-first low-pass system of Eq. 8.9 to obtain a low-pass precursor, which is then converted into the final high-pass ladder. Low-pass series inductors become high-pass series capacitors, and the low-pass shunt capacitor becomes a high-pass shunt inductor.

8.6.1 Design Example—Third-Order High-Pass Series-First

The design requirements are:

- $R_S = 148.5\,\text{k}\Omega$
- $R_L = 72.5\,\text{k}\Omega$
- $f_c = 444.4\,\text{kHz}$
- $\omega_c = 2.792247 \times 10^6\,\text{rad/s}$.

The passband gain established by the terminations is

$$20\log_{10}\left(\frac{72.5\,\mathrm{k}\Omega}{72.5\,\mathrm{k}\Omega + 148.5\,\mathrm{k}\Omega}\right) = -9.681085\,\mathrm{dB}.$$

Therefore the gain at the high-pass cut-on frequency is

$$20\log_{10}\left(\frac{1}{\sqrt{2}} * \frac{72.5\,\mathrm{k}\Omega}{72.5\,\mathrm{k}\Omega + 148.5\,\mathrm{k}\Omega}\right) = -12.691385\,\mathrm{dB},$$

or

$$-9.681085\,\mathrm{dB} - 3.010299\,\mathrm{dB} = -12.691385\,\mathrm{dB}.$$

To obtain the low-pass precursor for the high-pass realization, the series-first low-pass system of Eq. 8.9 is solved using the high-pass design requirements. A MATLAB implementation may be written as

```
format longEng

Rs = 148.5e3
Rl = 72.5e3
f = 444.4e3
w = 2*pi*f

syms L1 C2 L3

E0 = (w^0)*((Rs/Rl)+1)
E1 = (w^1)*(L1/Rl + C2*Rs + L3/Rl) ==
E0 * (cos(0)/sin((1*pi)/6))
E2 = (w^2)*(L1*C2 + C2*L3*(Rs/Rl)) ==
E0 * (cos(0)/sin((1*pi)/6))
E3 = (w^3)*(L1*C2*L3)/Rl == E0

result = solve(E1,E2,E3)

L1 = vpa(result.L1)
C2 = vpa(result.C2)
L3 = vpa(result.L3)
```

The complete candidate solution tuple output for the low-pass precursor solve used in this high-pass design example is provided in Appendix B, Sect. B.8.

For this low-pass precursor solve, the solver returns four candidate tuples: two fully real tuples and one complex-conjugate pair. The two fully real tuples are retained as valid passive realizations.

To provide a concrete component set for the high-pass transformation, one of the fully real tuples is selected as a representative low-pass precursor. The first fully real tuple is used here for the transformation and verification steps (Fig. 8.9):

$$L_1 = 53.1829636\,\mathrm{mH}, \qquad C_2 = 7.351465\,\mathrm{pF}, \qquad L_3 = 25.96474656\,\mathrm{mH}.$$

The corresponding high-pass elements are

$$2.411681\,\mathrm{pF} = \frac{1}{(2.792247 \times 10^6)^2(53.1829636\,\mathrm{mH})},$$

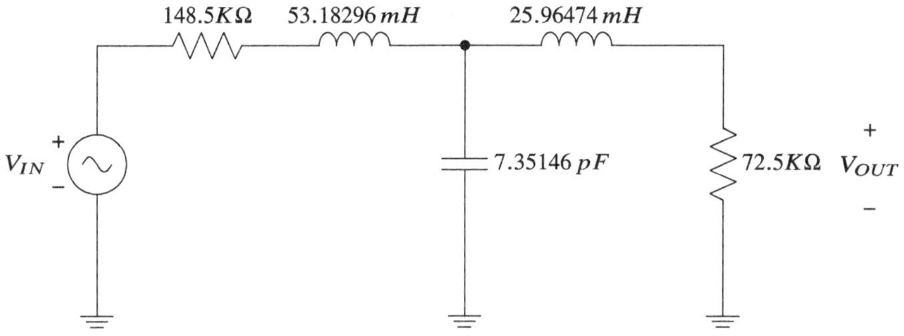

Fig. 8.9 Representative series-first low-pass precursor

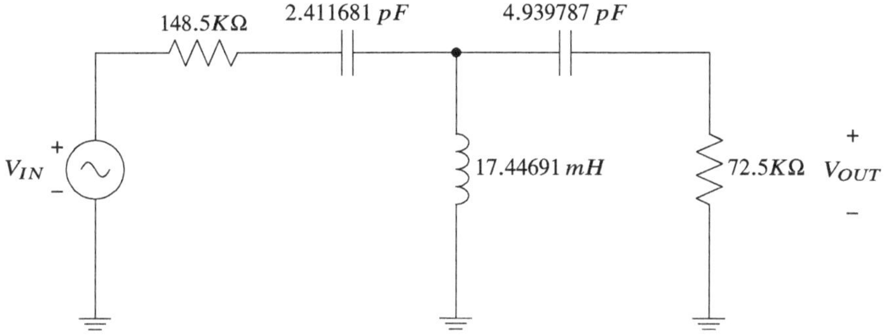

Fig. 8.10 Representative series-first high-pass solution

for the first series capacitor,

$$17.44691\ \text{mH} = \frac{1}{(2.792247 \times 10^6)^2(7.351465\ \text{pF})},$$

for the shunt inductor, and

$$4.939787\ \text{pF} = \frac{1}{(2.792247 \times 10^6)^2(25.96474656\ \text{mH})},$$

for the second series capacitor (Fig. 8.10).

8.7 Conclusion

The third-order case is the first higher odd-order synthesis chapter, and it shows clearly how Butterworth symmetry is projected into the realized ladder denominator. The third-degree term matches the constant term, while the second- and first-degree terms form a paired relationship. This is the circuit-level expression of the coefficient symmetry developed earlier in the monograph.

The chapter also shows that, at third order, the solution structure becomes richer than in the lower-order cases: multiple real realizations may coexist with complex-conjugate candidate tuples, and the time-constant identities begin to exhibit the finite trigonometric structure later collected in Chap. 12.

Accordingly, Chap. 9 turns to the fourth-order case, where the even-order symmetry pattern becomes more pronounced: the outer degrees remain paired, the odd interior degrees remain paired, and the second-degree term stands alone.

4th-Order Doubly-Terminated Passive Butterworth Filter Network Synthesis

9

Fourth order is where symmetry keeps its pattern but loses its simplicity.

Chapter 8 showed how odd-order symmetry is projected into a richer realized denominator structure: the outer degrees pair with the constant term, the inner degrees pair with one another, and the solution set expands to include both multiple real realizations and non-real candidate tuples. The present chapter extends the same synthesis framework to fourth order, where the even-order pattern becomes more pronounced.

As in the preceding chapters, the method itself does not change. Each degree-k frequency-dependent term of the realized ladder denominator is set equal, at cutoff, to the frequency-independent term multiplied by the corresponding normalized Butterworth coefficient. What changes at fourth order is the structure of the denominator and the resulting pattern of degree relationships. Here, the outermost degree remains paired with the constant term, the odd interior degrees remain paired, and the second-degree term stands alone. This is the clearest higher-order example yet of even-order asymmetry.

The Butterworth response used for coefficient matching traces to Butterworth's original maximally flat filter response [1]. Passive ladder-filter synthesis and classical analog-filter design are treated broadly in the established filter literature [2–5], with practical electronics context also given by Horowitz and Hill [6]. The particular time-constant matching method developed here, however, is the author's construction and is therefore derived directly rather than cited to an external source.

K. M. P. Ryan, *Polynomial Symmetry in Butterworth Filter Synthesis*,
Synthesis Lectures on Engineering, Science, and Technology,
https://doi.org/10.1007/978-3-032-33468-8_9

9.1 Even Symmetry

As with the second-order case, fourth-order doubly-terminated passive Butterworth filters are even-order and therefore topologically constrained. A shunt-first circuit topology is used when $R_S > R_L$, and a series-first circuit topology is used when $R_S < R_L$. When $R_S = R_L$, either topology may be used.

The fourth-order synthesis problem produces a larger candidate solution set than the lower-order cases. In the present work, fully real solutions are retained as valid passive realizations. Solutions containing imaginary components may still be algebraically informative, but they are not treated as exact passive filter designs.

Immediately before the fifth-order transition, it is useful to make the fourth-order factor relationships explicit. Table 9.1 reproduces the fourth-order product-factor structure in chapter context, while Appendix A.1 collects the order-by-order tables together for direct comparison. In the fourth-order case, all four factor lines remain part of a single correlated family: the lower two lines are generated by inversion and scaling of the upper two lines, and no secondary correlated factor family has yet appeared.

The factors in Table 9.1 are multiplied cumulatively to form the normalized fourth-order denominator polynomial. Written as a denominator function, this gives

$$\begin{aligned} D_4(s) = s^4 &+ (\text{Line 1})s^3 + (\text{Line 1})(\text{Line 2})s^2 \\ &+ (\text{Line 1})(\text{Line 2})(\text{Line 3})s \\ &+ (\text{Line 1})(\text{Line 2})(\text{Line 3})(\text{Line 4}), \end{aligned}$$

where any equivalent entry from the corresponding table line may be substituted. Using the numerical factors,

$$\begin{aligned} D_4(s) = s^4 &+ (2.613125)s^3 + (2.613125)(1.306562)s^2 \\ &+ (2.613125)(1.306562)(0.7653668)s \\ &+ (2.613125)(1.306562)(0.7653668)(0.3826834). \end{aligned}$$

The factors in Table 9.1 therefore establish the "before" case for the transition developed in the next chapter. Fourth order still admits a single correlated factor family. Fifth order introduces the first secondary correlated factor family, and sixth order makes that secondary family visible as a reciprocal pair.

Table 9.1 Fourth-order Butterworth polynomial factors with respect to Eq. 2.1 (Variations 1–4), and derivations (Variations 5–7)

Line	Factor	Variation 1	Variation 2	Variation 3	Variation 4	Variation 5	Variation 6	Variation 7
1	2.613125	$\left(\frac{\cos(0)}{\sin(\frac{1}{8}\pi)}\right)$	$2\left(\frac{\cos(\frac{1}{8}\pi)}{\sin(\frac{2}{8}\pi)}\right)$	$\frac{2}{\left(\frac{\cos(\frac{2}{8}\pi)}{\sin(\frac{3}{8}\pi)}\right)}$	$\frac{1}{\left(\frac{\cos(\frac{3}{8}\pi)}{\sin(\frac{4}{8}\pi)}\right)}$	$\left(\sqrt{2}\left(\sqrt{2+\sqrt{2}}\right)\right)$	$\left(\frac{2}{\left(\sqrt{2-\sqrt{2}}\right)}\right)$	$\csc(\frac{\pi}{8})$
2	1.306562	$\frac{\left(\frac{\cos(0)}{\sin(\frac{1}{8}\pi)}\right)}{2}$	$\left(\frac{\cos(\frac{1}{8}\pi)}{\sin(\frac{2}{8}\pi)}\right)$	$\frac{1}{\left(\frac{\cos(\frac{2}{8}\pi)}{\sin(\frac{3}{8}\pi)}\right)}$	$\frac{1}{2\left(\frac{\cos(\frac{3}{8}\pi)}{\sin(\frac{4}{8}\pi)}\right)}$	$\frac{\left(\sqrt{2}\left(\sqrt{2+\sqrt{2}}\right)\right)}{2}$	$\left(\frac{1}{\left(\sqrt{2-\sqrt{2}}\right)}\right)$	$\frac{\csc(\frac{\pi}{8})}{2}$
3	0.7653668	$\frac{2}{\left(\frac{\cos(0)}{\sin(\frac{1}{8}\pi)}\right)}$	$\frac{1}{\left(\frac{\cos(\frac{1}{8}\pi)}{\sin(\frac{2}{8}\pi)}\right)}$	$\left(\frac{\cos(\frac{2}{8}\pi)}{\sin(\frac{3}{8}\pi)}\right)$	$2\left(\frac{\cos(\frac{3}{8}\pi)}{\sin(\frac{4}{8}\pi)}\right)$	$\frac{2}{\left(\sqrt{2}\left(\sqrt{2+\sqrt{2}}\right)\right)}$	$\left(\left(\sqrt{2-\sqrt{2}}\right)\right)$	$\frac{2}{\csc(\frac{\pi}{8})}$
4	0.3826834	$\frac{1}{\left(\frac{\cos(0)}{\sin(\frac{1}{8}\pi)}\right)}$	$\frac{1}{2\left(\frac{\cos(\frac{1}{8}\pi)}{\sin(\frac{2}{8}\pi)}\right)}$	$\frac{\left(\frac{\cos(\frac{2}{8}\pi)}{\sin(\frac{3}{8}\pi)}\right)}{2}$	$\left(\frac{\cos(\frac{3}{8}\pi)}{\sin(\frac{4}{8}\pi)}\right)$	$\frac{1}{\left(\sqrt{2}\left(\sqrt{2+\sqrt{2}}\right)\right)}$	$\frac{\left(\left(\sqrt{2-\sqrt{2}}\right)\right)}{2}$	$\frac{1}{\csc(\frac{\pi}{8})}$

9.2 Low-Pass Shunt-First

The transfer function for the fourth-order shunt-first doubly-terminated low-pass filter in Fig. 9.1 is

$$\frac{V_{OUT}}{V_{IN}} = \frac{1}{\begin{aligned}&\omega^4 C_1 L_2 C_3 L_4 \frac{R_S}{R_L} \\ &+ \omega^3 \left(C_1 L_2 C_3 R_S + L_2 C_3 L_4 \frac{1}{R_L} \right) \\ &+ \omega^2 \left(C_1 L_2 \frac{R_S}{R_L} + C_1 L_4 \frac{R_S}{R_L} + C_3 L_2 + C_3 L_4 \frac{R_S}{R_L} \right) \\ &+ \omega \left(C_1 R_S + \frac{L_2}{R_L} + C_3 R_S + \frac{L_4}{R_L} \right) + \left(\frac{R_S}{R_L} + 1 \right)\end{aligned}} \tag{9.1}$$

The corresponding normalized Butterworth transfer function using the product formula of Eq. 2.1 is

$$\frac{V_{OUT}}{V_{IN}} = \frac{1}{\begin{aligned}&s^4 \left(\left(\frac{\cos(0)}{\sin(\frac{1}{8}\pi)} \right) \left(\frac{\cos(\frac{1}{8}\pi)}{\sin(\frac{2}{8}\pi)} \right) \left(\frac{\cos(\frac{2}{8}\pi)}{\sin(\frac{3}{8}\pi)} \right) \left(\frac{\cos(\frac{3}{8}\pi)}{\sin(\frac{4}{8}\pi)} \right) \right) \\ &+ s^3 \left(\left(\frac{\cos(0)}{\sin(\frac{1}{8}\pi)} \right) \left(\frac{\cos(\frac{1}{8}\pi)}{\sin(\frac{2}{8}\pi)} \right) \left(\frac{\cos(\frac{2}{8}\pi)}{\sin(\frac{3}{8}\pi)} \right) \right) \\ &+ s^2 \left(\left(\frac{\cos(0)}{\sin(\frac{1}{8}\pi)} \right) \left(\frac{\cos(\frac{1}{8}\pi)}{\sin(\frac{2}{8}\pi)} \right) \right) + s \left(\frac{\cos(0)}{\sin(\frac{1}{8}\pi)} \right) + 1\end{aligned}} \tag{9.2}$$

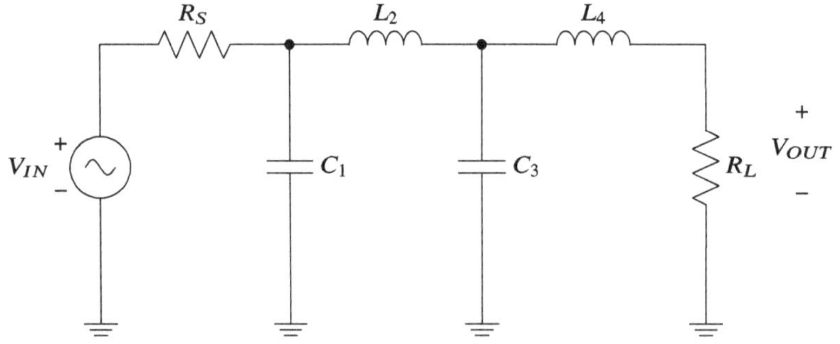

Fig. 9.1 Fourth-order shunt-first doubly-terminated low-pass Butterworth filter

The fourth-degree term of the realized denominator,

$$\omega^4 C_1 L_2 C_3 L_4 \frac{R_S}{R_L},$$

is associated with the fourth-degree Butterworth term

$$s^4 \left(\left(\frac{\cos(0)}{\sin(\frac{1}{8}\pi)} \right) \left(\frac{\cos(\frac{1}{8}\pi)}{\sin(\frac{2}{8}\pi)} \right) \left(\frac{\cos(\frac{2}{8}\pi)}{\sin(\frac{3}{8}\pi)} \right) \left(\frac{\cos(\frac{3}{8}\pi)}{\sin(\frac{4}{8}\pi)} \right) \right).$$

The third-degree term of the realized denominator,

$$\omega^3 \left(C_1 L_2 C_3 R_S + L_2 C_3 L_4 \frac{1}{R_L} \right),$$

is associated with the third-degree Butterworth term

$$s^3 \left(\left(\frac{\cos(0)}{\sin(\frac{1}{8}\pi)} \right) \left(\frac{\cos(\frac{1}{8}\pi)}{\sin(\frac{2}{8}\pi)} \right) \left(\frac{\cos(\frac{2}{8}\pi)}{\sin(\frac{3}{8}\pi)} \right) \right).$$

The second-degree term of the realized denominator,

$$\omega^2 \left(C_1 L_2 \frac{R_S}{R_L} + C_1 L_4 \frac{R_S}{R_L} + C_3 L_2 + C_3 L_4 \frac{R_S}{R_L} \right),$$

is associated with the second-degree Butterworth term

$$s^2 \left(\left(\frac{\cos(0)}{\sin(\frac{1}{8}\pi)} \right) \left(\frac{\cos(\frac{1}{8}\pi)}{\sin(\frac{2}{8}\pi)} \right) \right).$$

The first-degree term of the realized denominator,

$$\omega \left(C_1 R_S + \frac{L_2}{R_L} + C_3 R_S + \frac{L_4}{R_L} \right),$$

is associated with the first-degree Butterworth term

$$s \left(\frac{\cos(0)}{\sin(\frac{1}{8}\pi)} \right).$$

Applying the transformation rule yields

$$
\begin{aligned}
AC_{aspect} &= (DC_{aspect})\,(Butterworth_{coef}) \\
\omega^4 C_1 L_2 C_3 L_4 \frac{R_S}{R_L} &= \left(\frac{R_S}{R_L}+1\right)\left(\left(\frac{\cos(0)}{\sin(\frac{1}{8}\pi)}\right)\left(\frac{\cos(\frac{1}{8}\pi)}{\sin(\frac{2}{8}\pi)}\right)\left(\frac{\cos(\frac{2}{8}\pi)}{\sin(\frac{3}{8}\pi)}\right)\left(\frac{\cos(\frac{3}{8}\pi)}{\sin(\frac{4}{8}\pi)}\right)\right) \\
\omega^3\left(C_1 L_2 C_3 R_S + L_2 C_3 L_4 \frac{1}{R_L}\right) &= \left(\frac{R_S}{R_L}+1\right)\left(\left(\frac{\cos(0)}{\sin(\frac{1}{8}\pi)}\right)\left(\frac{\cos(\frac{1}{8}\pi)}{\sin(\frac{2}{8}\pi)}\right)\left(\frac{\cos(\frac{2}{8}\pi)}{\sin(\frac{3}{8}\pi)}\right)\right) \\
\omega^2\left(C_1 L_2 \frac{R_S}{R_L} + C_1 L_4 \frac{R_S}{R_L} + C_3 L_2 + C_3 L_4 \frac{R_S}{R_L}\right) &= \left(\frac{R_S}{R_L}+1\right)\left(\left(\frac{\cos(0)}{\sin(\frac{1}{8}\pi)}\right)\left(\frac{\cos(\frac{1}{8}\pi)}{\sin(\frac{2}{8}\pi)}\right)\right) \\
\omega\left(C_1 R_S + \frac{L_2}{R_L} + C_3 R_S + \frac{L_4}{R_L}\right) &= \left(\frac{R_S}{R_L}+1\right)\left(\frac{\cos(0)}{\sin(\frac{1}{8}\pi)}\right)
\end{aligned}
\tag{9.3}
$$

The system of Eq. 9.3 simplifies to

$$
\omega^4 C_1 L_2 C_3 L_4 \frac{R_S}{R_L} = \left(\frac{R_S}{R_L}+1\right), \tag{9.4}
$$

$$
\begin{aligned}
\omega^3\left(C_1 L_2 C_3 R_S + L_2 C_3 L_4 \frac{1}{R_L}\right) &= \left(\sqrt{2}\sqrt{2+\sqrt{2}}\right)\left(\frac{R_S}{R_L}+1\right) \\
&= \left(\frac{2}{\sqrt{2-\sqrt{2}}}\right)\left(\frac{R_S}{R_L}+1\right) \\
&= \left(\csc\left(\frac{\pi}{8}\right)\right)\left(\frac{R_S}{R_L}+1\right),
\end{aligned}
\tag{9.5}
$$

$$
\begin{aligned}
\omega^2\left(C_1 L_2 \frac{R_S}{R_L} + C_1 L_4 \frac{R_S}{R_L} + C_3 L_2 + C_3 L_4 \frac{R_S}{R_L}\right) &= \left(2+\sqrt{2}\right)\left(\frac{R_S}{R_L}+1\right) \\
&= \left(\frac{2}{2-\sqrt{2}}\right)\left(\frac{R_S}{R_L}+1\right) \\
&= \left(\frac{\csc(\frac{\pi}{8})^2}{2}\right)\left(\frac{R_S}{R_L}+1\right),
\end{aligned}
\tag{9.6}
$$

and

$$
\begin{aligned}
\omega\left(C_1 R_S + \frac{L_2}{R_L} + C_3 R_S + \frac{L_4}{R_L}\right) &= \left(\sqrt{2}\sqrt{2+\sqrt{2}}\right)\left(\frac{R_S}{R_L}+1\right) \\
&= \left(\frac{2}{\sqrt{2-\sqrt{2}}}\right)\left(\frac{R_S}{R_L}+1\right) \\
&= \left(\csc\left(\frac{\pi}{8}\right)\right)\left(\frac{R_S}{R_L}+1\right).
\end{aligned}
\tag{9.7}
$$

The fourth-degree term and the constant term are equal. The third- and first-degree terms are equal. The second-degree term stands alone. This is precisely the even-order pattern anticipated earlier.

Since there are four equations and four unknown reactive element values, the fourth-order system can produce multiple candidate solution tuples. In practice, the real-valued tuples are retained as exact passive realizations.

9.2.1 Design Example—Fourth-Order Low-Pass Shunt-First

The design requirements are:

- $R_S = 4.83\,\text{k}\Omega$
- $R_L = 4.72\,\text{k}\Omega$
- $f_c = 32.7\,\text{kHz}$
- $\omega_c = 205.4601 \times 10^3\,\text{rad/s}$.

The gain at DC, or throughout the passband as established by the terminations, is

$$20\log_{10}\left(\frac{4.72\,\text{k}\Omega}{4.83\,\text{k}\Omega + 4.72\,\text{k}\Omega}\right) = -6.121227\,\text{dB}.$$

Therefore the gain at cutoff is

$$20\log_{10}\left(\frac{1}{\sqrt{2}} * \frac{4.72\,\text{k}\Omega}{4.83\,\text{k}\Omega + 4.72\,\text{k}\Omega}\right) = -9.131527\,\text{dB},$$

or

$$-6.121227\,\text{dB} - 3.010299\,\text{dB} = -9.131526\,\text{dB}.$$

A MATLAB implementation of the system of equations in Eq. 9.3 may be written as

```
format shortEng

Rs=4.83e3
Rl=4.72e3
f=32.7e3
w=2*pi*f

syms C1 L2 C3 L4

E0 = ((Rs/Rl)+1)

E1 = w*(C1*Rs+L2/Rl+C3*Rs+L4/Rl) ==
     E0*(cos((0*pi)/8)/sin((1*pi)/8))

E2 = w^2*(C1*L2*Rs/Rl+C1*L4*Rs/Rl+
         C3*L4*Rs/Rl+C3*L2) ==
     E0*(cos((0*pi)/8)/sin((1*pi)/8))*
     (cos((1*pi)/8)/sin((2*pi)/8))


E3 = w^3*(C1*L2*C3*Rs+(L2*C3*L4)/Rl) ==
     E0*(cos((0*pi)/8)/sin((1*pi)/8))*
     (cos((1*pi)/8)/sin((2*pi)/8))*
     (cos((2*pi)/8)/sin((3*pi)/8))
```

```
E4 = w^4*C1*L2*C3*L4*(Rs/Rl) ==
     E0*(cos((0*pi)/8)/sin((1*pi)/8))*
     (cos((1*pi)/8)/sin((2*pi)/8))*
     (cos((2*pi)/8)/sin((3*pi)/8))*
     (cos((3*pi)/8)/sin((4*pi)/8))

result = solve(E1,E2,E3,E4)

C1 = vpa(result.C1)
L2 = vpa(result.L2)
C3 = vpa(result.C3)
L4 = vpa(result.L4)

M_C1 = abs(C1)
M_L2 = abs(L2)
M_C3 = abs(C3)
M_L4 = abs(L4)
```

The complete candidate solution tuple output for this design example is provided in Appendix B, Sect. B.9.

In this example, the solver returns eight candidate tuples: four fully real tuples and two complex-conjugate pairs. The four fully real tuples are retained as valid passive realizations.

To provide a concrete component set for verification, one of the fully real tuples is selected as a representative passive realization. The first fully real tuple is used here (Fig. 9.2):

$$C_1 = 1.147014408509\,\text{nF}, \qquad L_2 = 45.351245077\,\text{mH},$$

$$C_3 = 1.6105407809\,\text{nF}, \qquad L_4 = 13.24390151059\,\text{mH}.$$

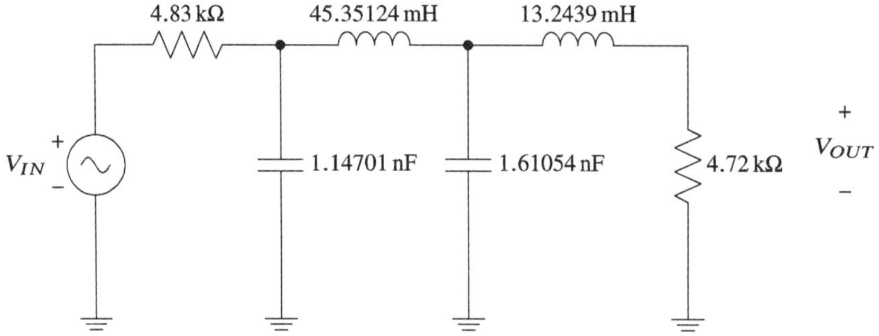

Fig. 9.2 First valid fourth-order shunt-first low-pass solution

9.3 Low-Pass Series-First

For $R_S < R_L$, the complementary fourth-order low-pass topology is series-first, as shown in Fig. 9.3. This is the appropriate even-order realization when the load resistance exceeds the source resistance.

The transfer function for the fourth-order series-first doubly-terminated low-pass filter in Fig. 9.3 is

$$\frac{V_{OUT}}{V_{IN}} = \frac{1}{\begin{array}{l}\omega^4 L_1 C_2 L_3 C_4 \\ + \omega^3 \left(L_1 C_2 L_3 \frac{1}{R_L} + C_2 L_3 C_4 R_S \right) \\ + \omega^2 \left(L_1 C_2 + C_2 L_3 \frac{R_S}{R_L} + L_1 C_4 + L_3 C_4 \right) \\ + \omega \left(\frac{L_1}{R_L} + C_2 R_S + \frac{L_3}{R_L} + C_4 R_S \right) + \left(\frac{R_S}{R_L} + 1 \right)\end{array}} \tag{9.8}$$

The corresponding normalized Butterworth transfer function using the product formula of Eq. 2.1 is

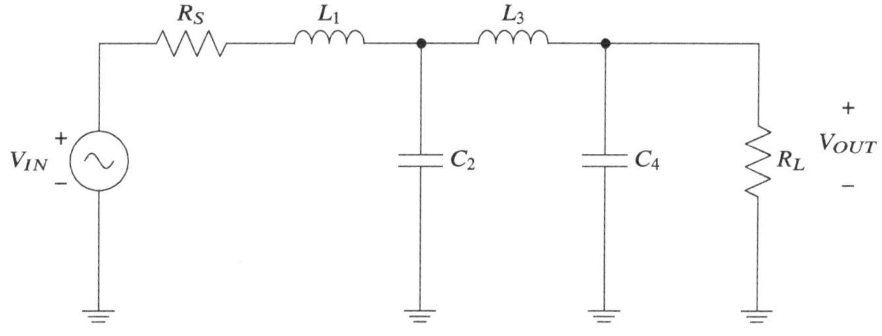

Fig. 9.3 Fourth-order series-first doubly-terminated low-pass Butterworth filter

$$\frac{V_{OUT}}{V_{IN}} = \frac{1}{s^4\left(\left(\frac{\cos(0)}{\sin(\frac{1}{8}\pi)}\right)\left(\frac{\cos(\frac{1}{8}\pi)}{\sin(\frac{2}{8}\pi)}\right)\left(\frac{\cos(\frac{2}{8}\pi)}{\sin(\frac{3}{8}\pi)}\right)\left(\frac{\cos(\frac{3}{8}\pi)}{\sin(\frac{4}{8}\pi)}\right)\right) + s^3\left(\left(\frac{\cos(0)}{\sin(\frac{1}{8}\pi)}\right)\left(\frac{\cos(\frac{1}{8}\pi)}{\sin(\frac{2}{8}\pi)}\right)\left(\frac{\cos(\frac{2}{8}\pi)}{\sin(\frac{3}{8}\pi)}\right)\right) + s^2\left(\left(\frac{\cos(0)}{\sin(\frac{1}{8}\pi)}\right)\left(\frac{\cos(\frac{1}{8}\pi)}{\sin(\frac{2}{8}\pi)}\right)\right) + s\left(\frac{\cos(0)}{\sin(\frac{1}{8}\pi)}\right) + 1} \tag{9.9}$$

The fourth-degree term of the realized denominator,

$$\omega^4 L_1 C_2 L_3 C_4,$$

is associated with the fourth-degree Butterworth term

$$s^4\left(\left(\frac{\cos(0)}{\sin(\frac{1}{8}\pi)}\right)\left(\frac{\cos(\frac{1}{8}\pi)}{\sin(\frac{2}{8}\pi)}\right)\left(\frac{\cos(\frac{2}{8}\pi)}{\sin(\frac{3}{8}\pi)}\right)\left(\frac{\cos(\frac{3}{8}\pi)}{\sin(\frac{4}{8}\pi)}\right)\right).$$

The third-degree term of the realized denominator,

$$\omega^3\left(L_1 C_2 L_3 \frac{1}{R_L} + C_2 L_3 C_4 R_S\right),$$

is associated with the third-degree Butterworth term

$$s^3\left(\left(\frac{\cos(0)}{\sin(\frac{1}{8}\pi)}\right)\left(\frac{\cos(\frac{1}{8}\pi)}{\sin(\frac{2}{8}\pi)}\right)\left(\frac{\cos(\frac{2}{8}\pi)}{\sin(\frac{3}{8}\pi)}\right)\right).$$

The second-degree term of the realized denominator,

$$\omega^2\left(L_1 C_2 + C_2 L_3 \frac{R_S}{R_L} + L_1 C_4 + L_3 C_4\right),$$

is associated with the second-degree Butterworth term

$$s^2\left(\left(\frac{\cos(0)}{\sin(\frac{1}{8}\pi)}\right)\left(\frac{\cos(\frac{1}{8}\pi)}{\sin(\frac{2}{8}\pi)}\right)\right).$$

The first-degree term of the realized denominator,

$$\omega\left(\frac{L_1}{R_L}+C_2R_S+\frac{L_3}{R_L}+C_4R_S\right),$$

is associated with the first-degree Butterworth term

$$s\left(\frac{\cos(0)}{\sin(\frac{1}{8}\pi)}\right).$$

Applying the transformation rule yields

$$\begin{aligned}
AC_{aspect} &= (DC_{aspect})\,(Butterworth_{coef})\\
\omega^4 L_1C_2L_3C_4 &= \left(\frac{R_S}{R_L}+1\right)\left(\left(\frac{\cos(0)}{\sin(\frac{1}{8}\pi)}\right)\left(\frac{\cos(\frac{1}{8}\pi)}{\sin(\frac{2}{8}\pi)}\right)\left(\frac{\cos(\frac{2}{8}\pi)}{\sin(\frac{3}{8}\pi)}\right)\left(\frac{\cos(\frac{3}{8}\pi)}{\sin(\frac{4}{8}\pi)}\right)\right)\\
\omega^3\left(L_1C_2L_3\frac{1}{R_L}+C_2L_3C_4R_S\right) &= \left(\frac{R_S}{R_L}+1\right)\left(\left(\frac{\cos(0)}{\sin(\frac{1}{8}\pi)}\right)\left(\frac{\cos(\frac{1}{8}\pi)}{\sin(\frac{2}{8}\pi)}\right)\left(\frac{\cos(\frac{2}{8}\pi)}{\sin(\frac{3}{8}\pi)}\right)\right)\\
\omega^2\left(L_1C_2+C_2L_3\frac{R_S}{R_L}+L_1C_4+L_3C_4\right) &= \left(\frac{R_S}{R_L}+1\right)\left(\left(\frac{\cos(0)}{\sin(\frac{1}{8}\pi)}\right)\left(\frac{\cos(\frac{1}{8}\pi)}{\sin(\frac{2}{8}\pi)}\right)\right)\\
\omega\left(\frac{L_1}{R_L}+C_2R_S+\frac{L_3}{R_L}+C_4R_S\right) &= \left(\frac{R_S}{R_L}+1\right)\left(\frac{\cos(0)}{\sin(\frac{1}{8}\pi)}\right)
\end{aligned} \tag{9.10}$$

The system of Eq. 9.10 simplifies to

$$\omega^4 L_1C_2L_3C_4 = \left(\frac{R_S}{R_L}+1\right), \tag{9.11}$$

$$\begin{aligned}
\omega^3\left(L_1C_2L_3\frac{1}{R_L}+C_2L_3C_4R_S\right) &= \left(\sqrt{2}\sqrt{2+\sqrt{2}}\right)\left(\frac{R_S}{R_L}+1\right)\\
&= \left(\frac{2}{\sqrt{2-\sqrt{2}}}\right)\left(\frac{R_S}{R_L}+1\right)\\
&= \left(\csc\left(\frac{\pi}{8}\right)\right)\left(\frac{R_S}{R_L}+1\right),
\end{aligned} \tag{9.12}$$

$$\begin{aligned}
\omega^2\left(L_1C_2+C_2L_3\frac{R_S}{R_L}+L_1C_4+L_3C_4\right) &= \left(2+\sqrt{2}\right)\left(\frac{R_S}{R_L}+1\right)\\
&= \left(\frac{2}{2-\sqrt{2}}\right)\left(\frac{R_S}{R_L}+1\right)\\
&= \left(\frac{\csc(\frac{\pi}{8})^2}{2}\right)\left(\frac{R_S}{R_L}+1\right),
\end{aligned} \tag{9.13}$$

and

$$\omega\left(\frac{L_1}{R_L}+C_2R_S+\frac{L_3}{R_L}+C_4R_S\right)=\left(\sqrt{2}\sqrt{2+\sqrt{2}}\right)\left(\frac{R_S}{R_L}+1\right)$$
$$=\left(\frac{2}{\sqrt{2-\sqrt{2}}}\right)\left(\frac{R_S}{R_L}+1\right) \tag{9.14}$$
$$=\left(\csc\left(\frac{\pi}{8}\right)\right)\left(\frac{R_S}{R_L}+1\right).$$

As in the shunt-first case, the fourth-degree term and the constant term are equal, the third- and first-degree terms are equal, and the second-degree term stands alone. This is the denominator-level manifestation of the even-order symmetry discussed earlier in Sect. 2.3.2.1.

Since there are four equations and four unknown reactive element values, the fourth-order series-first system can produce multiple candidate solution tuples. In practice, the real-valued tuples are retained as exact passive realizations.

9.3.1 Design Example—Fourth-Order Low-Pass Series-First

The design requirements are:

- $R_S = 13.8\,\text{k}\Omega$
- $R_L = 42.2\,\text{k}\Omega$
- $f_c = 300\,\text{kHz}$
- $\omega_c = 1.884955 \times 10^6\,\text{rad/s}$.

The gain at DC, or throughout the passband as established by the terminations, is

$$20\log_{10}\left(\frac{42.2\,\text{k}\Omega}{42.2\,\text{k}\Omega+13.8\,\text{k}\Omega}\right)=-2.457511\,\text{dB}.$$

Therefore the gain at cutoff is

$$20\log_{10}\left(\frac{1}{\sqrt{2}}*\frac{42.2\,\text{k}\Omega}{42.2\,\text{k}\Omega+13.8\,\text{k}\Omega}\right)=-5.467811\,\text{dB},$$

or

$$-2.457511\,\text{dB}-3.010299\,\text{dB}=-5.467811\,\text{dB}.$$

A MATLAB implementation of the system of equations in Eq. 9.10 may be written as

```
format shortEng

Rs = 13.8e3
Rl = 42.2e3
f = 300e3
w = 2*pi*f
```

```
syms L1 C2 L3 C4

E0 = w^0*((Rs/Rl)+1)

E1 = w^1*(L1/Rl+C2*Rs+L3/Rl+C4*Rs) == E0*(cos((0*pi)/8)/sin((1*pi)/8))

E2 = w^2*(L1*C2+L1*C4+L3*C4+L3*C2*Rs/Rl) == E0*(cos((0*pi)/8)/sin((1*pi)/8))*
                                            (cos((1*pi)/8)/sin((2*pi)/8))

E3 = w^3*(L1*C2*L3/Rl+(C2*L3*C4)*Rs) == E0*(cos((0*pi)/8)/sin((1*pi)/8))*
                                        (cos((1*pi)/8)/sin((2*pi)/8))*
                                        (cos((2*pi)/8)/sin((3*pi)/8))

E4 = w^4*L1*C2*L3*C4 == E0*(cos((0*pi)/8)/sin((1*pi)/8))*
                        (cos((1*pi)/8)/sin((2*pi)/8))*
                        (cos((2*pi)/8)/sin((3*pi)/8))*
                        (cos((3*pi)/8)/sin((4*pi)/8))


result = solve(E1,E2,E3,E4);

L1 = vpa(result.L1)
C2 = vpa(result.C2)
L3 = vpa(result.L3)
C4 = vpa(result.C4)

M_L1 = abs(L1)
M_C2 = abs(C2)
M_L3 = abs(L3)
M_C4 = abs(C4)
```

The complete candidate solution tuple output for this design example is provided in Appendix B, Sect. B.10.

In this example, the solver returns eight candidate tuples: four fully real tuples and two complex-conjugate pairs. The four fully real tuples are retained as valid passive realizations.

To provide a concrete component set for verification, one of the fully real tuples is selected as a representative passive realization. The third fully real tuple is used here (Fig. 9.4):

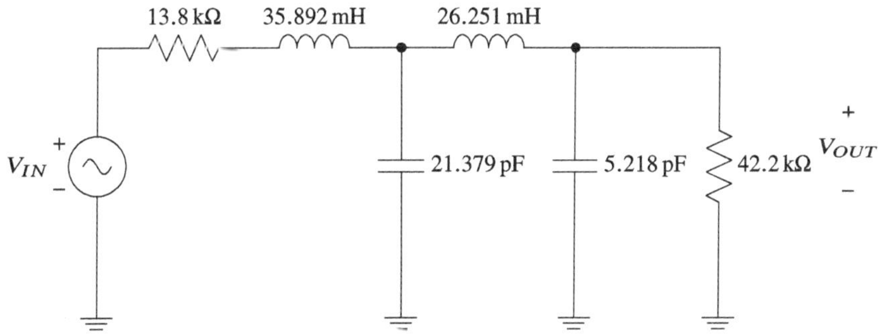

Fig. 9.4 Third valid fourth-order series-first low-pass solution

$$L_1 = 35.89213239\,\text{mH}, \qquad C_2 = 21.3791589\,\text{pF},$$
$$L_3 = 26.2517730237\,\text{mH}, \qquad C_4 = 5.2182221779\,\text{pF}.$$

9.4 Time Constants

Equations 6.8 and 6.17 define τ_C and τ_L, which are used throughout this section. For the fourth-order shunt-first low-pass filter, the first-degree portion of the system is

$$\omega\left(C_1 R_S + \frac{L_2}{R_L} + C_3 R_S + \frac{L_4}{R_L}\right) = \left(\frac{R_S}{R_L} + 1\right)\left(\sqrt{2}\sqrt{2+\sqrt{2}}\right),$$
$$\omega\left(C_1 \frac{R_S R_L}{R_S + R_L} + \frac{L_2}{R_S + R_L} + C_3 \frac{R_S R_L}{R_S + R_L} + \frac{L_4}{R_S + R_L}\right) = \left(\sqrt{2}\sqrt{2+\sqrt{2}}\right). \tag{9.15}$$

For the fourth-order series-first low-pass filter, the corresponding first-degree portion is

$$\omega\left(\frac{L_1}{R_L} + C_2 R_S + \frac{L_3}{R_L} + C_4 R_S\right) = \left(\frac{R_S}{R_L} + 1\right)\left(\sqrt{2}\sqrt{2+\sqrt{2}}\right),$$
$$\omega\left(\frac{L_1}{R_S + R_L} + C_2 \frac{R_S R_L}{R_S + R_L} + \frac{L_3}{R_S + R_L} + C_4 \frac{R_S R_L}{R_S + R_L}\right) = \left(\sqrt{2}\sqrt{2+\sqrt{2}}\right). \tag{9.16}$$

Thus, both fourth-order topologies resolve to

$$\omega\left(\tau_{C_1} + \tau_{L_2} + \tau_{C_3} + \tau_{L_4}\right) = \omega\left(\tau_{L_1} + \tau_{C_2} + \tau_{L_3} + \tau_{C_4}\right) = \left(\sqrt{2}\sqrt{2+\sqrt{2}}\right) = \frac{1}{\sin(\frac{1}{8}\pi)}.$$

Equivalently,

$$\begin{aligned}\omega\tau_{C_1} + \omega\tau_{L_2} + \omega\tau_{C_3} + \omega\tau_{L_4} &= \omega\tau_{L_1} + \omega\tau_{C_2} + \omega\tau_{L_3} + \omega\tau_{C_4} \\ &= \frac{1}{\sin(\frac{1}{8}\pi)}.\end{aligned} \tag{9.17}$$

If $R_S = R_L = 1$, then

$$\omega\tau_{C_1} = \omega\tau_{L_1} = \sin\left(\frac{1}{8}\pi\right),$$
$$\omega\tau_{L_2} = \omega\tau_{C_2} = \sin\left(\frac{3}{8}\pi\right),$$
$$\omega\tau_{C_3} = \omega\tau_{L_3} = \sin\left(\frac{5}{8}\pi\right),$$
$$\omega\tau_{L_4} = \omega\tau_{C_4} = \sin\left(\frac{7}{8}\pi\right),$$

so that

$$\sin\left(\frac{1}{8}\pi\right) + \sin\left(\frac{3}{8}\pi\right) + \sin\left(\frac{5}{8}\pi\right) + \sin\left(\frac{7}{8}\pi\right) = \frac{1}{\sin(\frac{1}{8}\pi)} = \csc\left(\frac{1}{8}\pi\right). \tag{9.18}$$

If $R_S \neq R_L$, then

$$\begin{aligned}
\omega\tau_{C_1} &\neq \omega\tau_{L_1} \neq \sin\left(\frac{1}{8}\pi\right),\\
\omega\tau_{L_2} &\neq \omega\tau_{C_2} \neq \sin\left(\frac{3}{8}\pi\right),\\
\omega\tau_{C_3} &\neq \omega\tau_{L_3} \neq \sin\left(\frac{5}{8}\pi\right),\\
\omega\tau_{L_4} &\neq \omega\tau_{C_4} \neq \sin\left(\frac{7}{8}\pi\right),
\end{aligned} \tag{9.19}$$

yet the summed identity of Eq. 9.17 remains valid.

This is the fourth-order instance of the finite trigonometric time-constant structure that is later collected in Chap. 12.

9.5 High-Pass Shunt-First

A fourth-order shunt-first high-pass filter is obtained by first synthesizing the corresponding fourth-order shunt-first low-pass ladder for the $R_S > R_L$ topology and then converting each reactive element using Eqs. 6.19 and 6.20. Therefore, the high-pass example is not a separate coefficient-matching derivation. It is a transformation example that uses the shunt-first low-pass system of Eq. 9.3 to obtain a low-pass precursor, which is then converted into the final high-pass ladder. Low-pass shunt capacitors become high-pass shunt inductors, and low-pass series inductors become high-pass series capacitors.

9.5.1 Design Example—Fourth-Order High-Pass Shunt-First

The design requirements are:

- $R_S = 47.38\,\text{k}\Omega$
- $R_L = 27.8\,\text{k}\Omega$
- $f_c = 2.54\,\text{MHz}$
- $\omega_c = 15.95929 \times 10^6\,\text{rad/s}$.

The passband gain established by the terminations is

$$20\log_{10}\left(\frac{27.8\,\text{k}\Omega}{47.38\,\text{k}\Omega + 27.8\,\text{k}\Omega}\right) = -8.641151\,\text{dB}.$$

Therefore the gain at the high-pass cut-on frequency is

$$20\log_{10}\left(\frac{1}{\sqrt{2}} * \frac{27.8\,\mathrm{k}\Omega}{47.38\,\mathrm{k}\Omega + 27.8\,\mathrm{k}\Omega}\right) = -11.651450\,\mathrm{dB},$$

or

$$-8.641151\,\mathrm{dB} - 3.010299\,\mathrm{dB} = -11.651450\,\mathrm{dB}.$$

A MATLAB implementation of the system of equations in Eq. 9.3 may be written as

```
format shortEng

Rs = 47.38e3
Rl = 27.8e3
f = 2.54e6
w = f*2*pi

syms C1 L2 C3 L4

E0 = ((Rs/Rl)+1)

E1 = w*(C1*Rs+L2/Rl+C3*Rs+L4/Rl) ==
 E0*(cos((0*pi)/8)/sin((1*pi)/8))

E2 = w^2*(C1*L2*Rs/Rl+C1*L4*Rs/Rl+C3*L4*Rs/Rl+C3*L2) ==
   E0*(cos((0*pi)/8)/sin((1*pi)/8))*
   (cos((1*pi)/8)/sin((2*pi)/8))


E3 = w^3*(C1*L2*C3*Rs+(L2*C3*L4)/Rl) ==
   E0*(cos((0*pi)/8)/sin((1*pi)/8))*
   (cos((1*pi)/8)/sin((2*pi)/8))*
   (cos((2*pi)/8)/sin((3*pi)/8))


E4 = w^4*E4 = w^4*C1*L2*C3*L4*(Rs/Rl) ==
   E0*(cos((0*pi)/8)/sin((1*pi)/8))*
   (cos((1*pi)/8)/sin((2*pi)/8))*
   (cos((2*pi)/8)/sin((3*pi)/8))*
   (cos((3*pi)/8)/sin((4*pi)/8))

result = solve(E1,E2,E3,E4)

C1 = vpa(result.C1)
L2 = vpa(result.L2)
C3 = vpa(result.C3)
L4 = vpa(result.L4)

M_C1 = abs(C1)
M_L2 = abs(L2)
M_C3 = abs(C3)
M_L4 = abs(L4)
```

The complete candidate solution tuple output for the low-pass precursor solve used in this high-pass design example is provided in Appendix B, Sect.B.11.

For this low-pass precursor solve, the solver returns eight candidate tuples: four fully real tuples and two complex-conjugate solution pairs. Each complex-conjugate pair also produces a magnitude tuple. These magnitude tuples are less accurate than the fully real

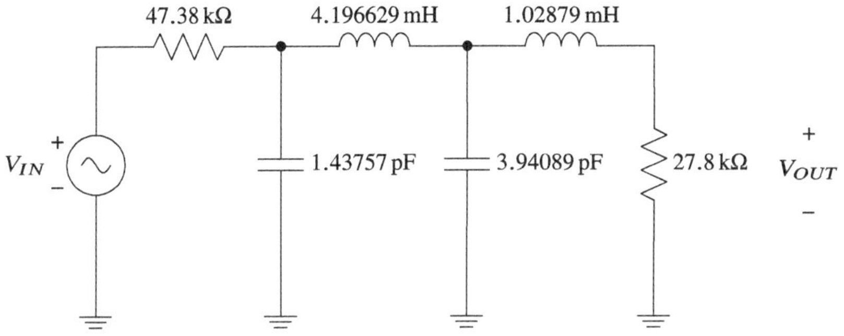

Fig. 9.5 Representative fourth-order shunt-first low-pass precursor for high-pass transformation

tuples and are not used as the representative passive realization. The four fully real tuples are retained as valid passive realizations.

To provide a concrete component set for the high-pass transformation, one of the fully real tuples is selected as a representative low-pass precursor. The second fully real tuple is used here for the transformation and verification steps (Fig. 9.5):

$$C_1 = 1.4375719413\,\text{pF}, \qquad L_2 = 4.1966299725\,\text{mH},$$

$$C_3 = 3.9408950289\,\text{pF}, \qquad L_4 = 1.0287925068\,\text{mH}.$$

Both low-pass shunt capacitors (C_1 and C_3) get transformed to high-pass shunt inductors (L_1 and L_3), and both low-pass series inductors (L_2 and L_4) get transformed to high-pass series capacitors (C_2 and C_4) using Equations 6.19 and 6.20 (Fig. 9.6).

The first high-pass shunt inductor, L_1, which is derived from C_1, becomes:

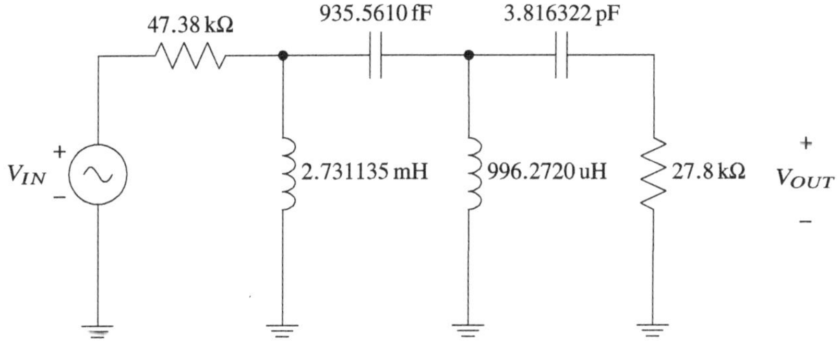

Fig. 9.6 Representative fourth-order shunt-first high-pass solution

$$2.731135\,\text{mH} = \frac{1}{(15.95929 \times 10^6 \frac{r}{s})^2 * (1.4375719413\,\text{pF})}$$

The first high-pass series capacitor, C_2, which is derived from L_2, becomes:

$$935.5610\,\text{fF} = \frac{1}{(15.95929 \times 10^6 \frac{r}{s})^2 * (4.1966299725\,\text{mH})}$$

The second high-pass shunt inductor, L_3, which is derived from C_3, becomes:

$$996.2720\,\text{uH} = \frac{1}{(15.95929 \times 10^6 \frac{r}{s})^2 * (3.9408950289\,\text{pF})}$$

The second high-pass series capacitor, C_4, which is derived from L_4, becomes:

$$3.816322\,\text{pF} = \frac{1}{(15.95929 \times 10^6 \frac{r}{s})^2 * (1.0287925068\,\text{mH})}$$

Producing a fourth-order shunt-first high-pass Butterworth filter.

9.6 High-Pass Series-First

A fourth-order series-first high-pass filter is obtained by first synthesizing the corresponding fourth-order series-first low-pass ladder for the $R_S < R_L$ topology and then converting each reactive element using Eqs. 6.19 and 6.20. Therefore, the high-pass example is not a separate coefficient-matching derivation. It is a transformation example that uses the series-first low-pass system of Eq. 9.10 to obtain a low-pass precursor, which is then converted into the final high-pass ladder. Low-pass series inductors become high-pass series capacitors, and low-pass shunt capacitors become high-pass shunt inductors.

9.6.1 Design Example—Fourth-Order High-Pass Series-First

The design requirements are:

- $R_S = 1.1\,\text{k}\Omega$
- $R_L = 1.11\,\text{k}\Omega$
- $f_c = 4.56\,\text{kHz}$
- $\omega_c = 28.65132 \times 10^3\,\text{rad/s}$.

The passband gain established by the terminations is

$$20\log_{10}\left(\frac{1.11\,\text{k}\Omega}{1.11\,\text{k}\Omega + 1.1\,\text{k}\Omega}\right) = -5.981385\,\text{dB}.$$

Therefore the gain at the high-pass cut-on frequency is

$$20 \log_{10}\left(\frac{1}{\sqrt{2}} * \frac{1.11\,\mathrm{k}\Omega}{1.11\,\mathrm{k}\Omega + 1.1\,\mathrm{k}\Omega}\right) = -8.991685\,\mathrm{dB},$$

or

$$-5.981385\,\mathrm{dB} - 3.010299\,\mathrm{dB} = -8.991685\,\mathrm{dB}.$$

A MATLAB implementation of the system of equations in Eq. 9.10 may be written as

```
format shortEng

Rs = 1.1e3
Rl = 1.11e3
f = 4.56e3
w = 2*pi*f

syms L1 C2 L3 C4

E0 = w^0*((Rs/Rl)+1)

E1 = w^1*(L1/Rl+C2*Rs+L3/Rl+C4*Rs) == E0*(cos((0*pi)/8)/sin((1*pi)/8))

E2 = w^2*(L1*C2+L1*C4+L3*C4+L3*C2*Rs/Rl) == E0*(cos((0*pi)/8)/sin((1*pi)/8))*
                                            (cos((1*pi)/8)/sin((2*pi)/8))

E3 = w^3*(L1*C2*L3/Rl+(C2*L3*C4)*Rs) == E0*(cos((0*pi)/8)/sin((1*pi)/8))*
                                        (cos((1*pi)/8)/sin((2*pi)/8))*
                                        (cos((2*pi)/8)/sin((3*pi)/8))

E4 = w^4*L1*C2*L3*C4 == E0*(cos((0*pi)/8)/sin((1*pi)/8))*
                        (cos((1*pi)/8)/sin((2*pi)/8))*
                        (cos((2*pi)/8)/sin((3*pi)/8))*
                        (cos((3*pi)/8)/sin((4*pi)/8))


result = solve(E1,E2,E3,E4);

L1 = vpa(result.L1)
C2 = vpa(result.C2)
L3 = vpa(result.L3)
C4 = vpa(result.C4)

M_L1 = abs(L1)
M_C2 = abs(C2)
M_L3 = abs(L3)
M_C4 = abs(C4)
```

The complete candidate solution tuple output for the low-pass precursor solve used in this high-pass design example is provided in Appendix B, Sect. B.12.

For this low-pass precursor solve, the solver returns eight candidate tuples: four fully real tuples and two complex-conjugate solution pairs. Each complex-conjugate pair also produces a magnitude tuple. These magnitude tuples are less accurate than the fully real tuples and are not used as the representative passive realization. The four fully real tuples are retained as valid passive realizations.

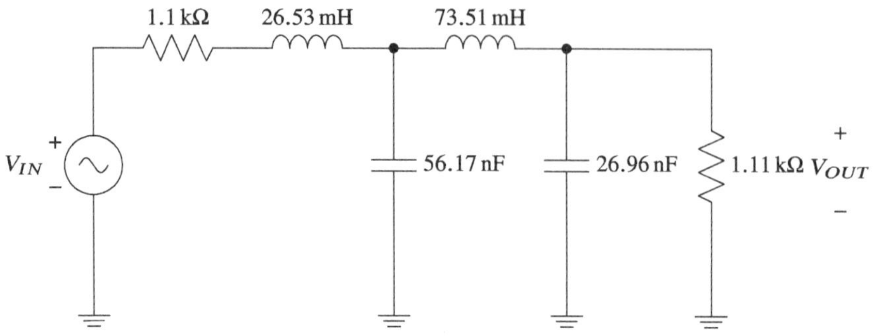

Fig. 9.7 Representative fourth-order series-first low-pass precursor for high-pass transformation

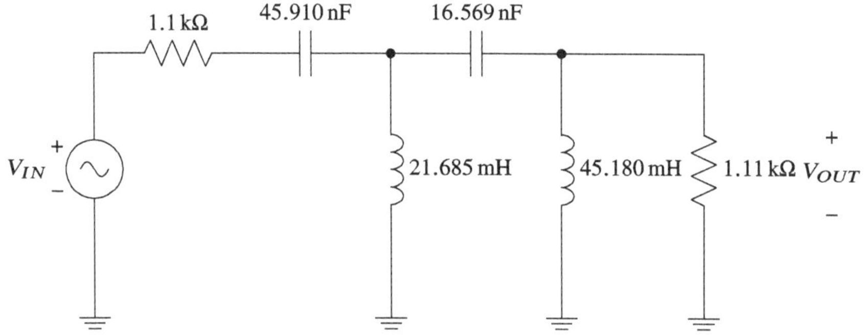

Fig. 9.8 Representative fourth-order series-first high-pass solution

To provide a concrete component set for the high-pass transformation, one of the fully real tuples is selected as a representative low-pass precursor. The third fully real tuple is used here for the transformation and verification steps (Fig. 9.7).

Both low-pass series inductors (L_1 and L_3) get transformed to high-pass series capacitors (C_1 and C_3), and both low-pass shunt capacitors (C_2 and C_4) get transformed to high-pass shunt inductors (L_2 and L_4) using Eqs. 6.19 and 6.20 (Fig. 9.8).

The first high-pass series capacitor, C_1, which is derived from L_1 becomes:

$$45.91021\,\text{nF} = \frac{1}{(28.65132 \times 10^3 \frac{rad}{s})^2 * (26.53391179\,\text{mH})}$$

The first high-pass shunt inductor, L_2, which is derived from C_2, becomes:

$$21.68595\,\text{mH} = \frac{1}{(28.65132 \times 10^3 \frac{rad}{s})^2 * (56.173577\,\text{nF})}$$

The second high-pass series capacitor, C_3, which is derived from L_3, becomes:

$$16.56964\,\text{nF} = \frac{1}{(28.65132 \times 10^3 \frac{rad}{s})^2 * (73.5186164\,\text{mH})}$$

The second high-pass shunt inductor, L_4, which is derived from C_4, becomes:

$$45.18047\,\text{mH} = \frac{1}{(28.65132 \times 10^3 \frac{rad}{s})^2 * (26.962475\,\text{nF})}$$

Producing a fourth-order series-first high-pass Butterworth filter.

9.7 Conclusion

The fourth-order case extends the exact synthesis method without changing its principle. The denominator is still matched degree by degree to the corresponding Butterworth structure, but the pattern of symmetry now takes its distinctly even-order form: fourth degree paired with the constant term, third degree paired with first degree, and second degree standing alone.

This chapter also reinforces the emerging time-constant pattern that later becomes one of the mathematical identities of the monograph. In that sense, fourth order is not merely another worked example; it is part of the evidence for the broader finite-series structure developed later.

The next chapter extends the same synthesis framework to fifth order, where the first secondary correlated factor family appears in transparent form.

References

1. Butterworth, S. (1930). On the theory of filter amplifiers. *Experimental Wireless and the Wireless Engineer, 7*, 536–541.
2. Zverev, A. I. (1967). *Handbook of filter synthesis*. Wiley.
3. Van Valkenburg, M. E. (1982). *Analog filter design*. Holt, Rinehart and Winston.
4. Schaumann, R., & Van Valkenburg, M. E. (2001). *Design of analog filters*. Oxford University Press.
5. Huelsman, L. P. (1993). *Active and passive analog filter design*. McGraw-Hill.
6. Horowitz, P., & Hill, W. (2015). *The art of electronics* (3rd ed.). Cambridge University Press.

5th-Order Doubly-Terminated Passive Butterworth Filter Network Synthesis

10

Fifth order is where the hidden transition first appears.

Chapter 9 showed the fourth-order case immediately before the product-factor-family transition. At fourth order, the product factors remain part of a single correlated family: the outer factors and inner factors may be expressed through inversion and scaling relationships, and no separate factor family is required. The fifth-order case preserves odd-order denominator symmetry, but it also introduces the first structural indication that this single-family pattern will not continue indefinitely.

As in the preceding chapters, the synthesis method itself does not change. Each degree-k term of the realized ladder denominator is set equal, at cutoff, to the frequency-independent term multiplied by the corresponding normalized Butterworth coefficient. What changes at fifth order is the product-factor-level interpretation of those coefficients.

The Butterworth coefficient structure used here traces to Butterworth's original maximally flat response [1]. Passive ladder-filter synthesis and classical analog-filter design are treated broadly in the established filter literature [3–6], with practical electronics context also given by Horowitz and Hill [2]. The product-factor-family interpretation and time-constant matching used in this chapter are developed as part of the present construction.

10.1 Odd Symmetry

Fifth-order doubly-terminated passive Butterworth ladder filters are odd-order and structurally symmetric. Accordingly, either a shunt-first topology or a series-first topology may be used regardless of the relative values of R_S and R_L.

K. M. P. Ryan, *Polynomial Symmetry in Butterworth Filter Synthesis*,
Synthesis Lectures on Engineering, Science, and Technology,
https://doi.org/10.1007/978-3-032-33468-8_10

In the fifth-order case, the mapped system contains five unknown reactive element values. The resulting nonlinear system may produce multiple candidate solution tuples, and the solution-branch structure becomes more prominent than in the lower-order examples. As in the lower-order cases, real-valued tuples are retained as valid passive realizations, while complex-valued tuples are not treated as physical filter designs in the present work. The increased solution complexity observed at fifth order is treated here as structurally significant, but not as proof that the synthesis method itself fails.

The fifth-order case is also the first order where the product-factor table is no longer fully described by the primary correlated factor family. Line 3 of Table 10.1 marks the first appearance of a secondary correlated factor family. In fifth order this secondary factor evaluates to unity, so it does not visibly disturb the simplified normalized polynomial. Structurally, however, it prepares the transition to sixth order, where the secondary family becomes visible as a reciprocal pair. This makes fifth order the first case where the product-factor structure and the observed solution-branch complexity begin to change at the same order.

Table 10.1 gives the fifth-order Butterworth product factors in chapter context. The complete order-by-order comparison remains collected in Appendix A.

Variations 6–9 are not separate fifth-order product-factor structures; rather, they are algebraically equivalent ways of writing the same underlying factor relationships. Consequently, any equivalent entry from the corresponding table line may be substituted when forming the denominator.

The factors in Table 10.1 are multiplied cumulatively to form the normalized fifth-order denominator polynomial. Written as a denominator function, this gives

$$
\begin{aligned}
D_5(s) = s^5 &+ (\text{Line }1)s^4 + (\text{Line }1)(\text{Line }2)s^3 \\
&+ (\text{Line }1)(\text{Line }2)(\text{Line }3)s^2 \\
&+ (\text{Line }1)(\text{Line }2)(\text{Line }3)(\text{Line }4)s \\
&+ (\text{Line }1)(\text{Line }2)(\text{Line }3)(\text{Line }4)(\text{Line }5),
\end{aligned}
$$

where any equivalent entry from the corresponding table line may be substituted. Using the numerical factors,

$$
\begin{aligned}
D_5(s) = s^5 &+ (3.236067)s^4 + (3.236067)(1.6180339)s^3 \\
&+ (3.236067)(1.6180339)(1)s^2 \\
&+ (3.236067)(1.6180339)(1)(0.6180339)s \\
&+ (3.236067)(1.6180339)(1)(0.6180339)(0.3090169).
\end{aligned}
$$

Table 10.1 Fifth-order Butterworth polynomial factors with respect to Eq. (2.1) (variations 1–5), and derivations (Variations 6–9)

Line	Factor	Variation 1	Variation 2	Variation 3	Variation 4	Variation 5	Variation 6	Variation 7	Variation 8	Variation 9
1	3.236067	$\left(\frac{\cos(0)}{\sin\left(\frac{1}{10}\pi\right)}\right)$	$2\left(\frac{\cos\left(\frac{1}{10}\pi\right)}{\sin\left(\frac{2}{10}\pi\right)}\right)$		$\frac{2}{\left(\frac{\cos\left(\frac{3}{10}\pi\right)}{\sin\left(\frac{4}{10}\pi\right)}\right)}$	$\frac{1}{\left(\frac{\cos\left(\frac{4}{10}\pi\right)}{\sin\left(\frac{5}{10}\pi\right)}\right)}$	$\left(\sqrt{5}+1\right)$	$\left(\frac{4}{(\sqrt{5}-1)}\right)$	$\csc\left(\frac{\pi}{10}\right)$	2Φ
2	1.6180339	$\frac{\left(\frac{\cos(0)}{\sin\left(\frac{1}{10}\pi\right)}\right)}{2}$	$\left(\frac{\cos\left(\frac{1}{10}\pi\right)}{\sin\left(\frac{2}{10}\pi\right)}\right)$		$\frac{1}{\left(\frac{\cos\left(\frac{3}{10}\pi\right)}{\sin\left(\frac{4}{10}\pi\right)}\right)}$	$\frac{1}{2\left(\frac{\cos\left(\frac{4}{10}\pi\right)}{\sin\left(\frac{5}{10}\pi\right)}\right)}$	$\left(\frac{\sqrt{5}+1}{2}\right)$	$\left(\frac{2}{(\sqrt{5}-1)}\right)$	$\frac{\csc\left(\frac{\pi}{10}\right)}{2}$	Φ
3	1			$\left(\frac{\cos\left(\frac{2}{10}\pi\right)}{\sin\left(\frac{3}{10}\pi\right)}\right)$						1
4	0.6180339	$\frac{2}{\left(\frac{\cos(0)}{\sin\left(\frac{1}{10}\pi\right)}\right)}$	$\frac{1}{\left(\frac{\cos\left(\frac{1}{10}\pi\right)}{\sin\left(\frac{2}{10}\pi\right)}\right)}$		$\left(\frac{\cos\left(\frac{3}{10}\pi\right)}{\sin\left(\frac{4}{10}\pi\right)}\right)$	$2\left(\frac{\cos\left(\frac{4}{10}\pi\right)}{\sin\left(\frac{5}{10}\pi\right)}\right)$	$\left(\frac{2}{\sqrt{5}+1}\right)$	$\left(\frac{\sqrt{5}-1}{2}\right)$	$\frac{2}{\csc\left(\frac{\pi}{10}\right)}$	$\frac{1}{\Phi}$
5	0.3090169	$\frac{1}{\left(\frac{\cos(0)}{\sin\left(\frac{1}{10}\pi\right)}\right)}$	$\frac{1}{2\left(\frac{\cos\left(\frac{1}{10}\pi\right)}{\sin\left(\frac{2}{10}\pi\right)}\right)}$		$\frac{\left(\frac{\cos\left(\frac{3}{10}\pi\right)}{\sin\left(\frac{4}{10}\pi\right)}\right)}{2}$	$\left(\frac{\cos\left(\frac{4}{10}\pi\right)}{\sin\left(\frac{5}{10}\pi\right)}\right)$	$\left(\frac{1}{\sqrt{5}+1}\right)$	$\left(\frac{\sqrt{5}-1}{4}\right)$	$\frac{1}{\csc\left(\frac{\pi}{10}\right)}$	$\frac{1}{2\Phi}$

The Line 3 value is numerically transparent in the fifth-order case, but it marks the first appearance of the secondary correlated factor family that becomes visible as a reciprocal pair in sixth order.

10.2 Low-Pass Shunt-First

The transfer function for the fifth-order shunt-first doubly-terminated low-pass Butterworth filter in Fig. 10.1 is

$$\frac{V_{OUT}}{V_{IN}} = \frac{1}{\begin{array}{l} \omega^5 C_1 L_2 C_3 L_4 C_5 R_S \\ + \omega^4 \left(C_1 L_2 C_3 L_4 \frac{R_S}{R_L} + L_2 C_3 L_4 C_5 \right) \\ + \omega^3 \left(C_1 L_2 C_3 R_S + C_1 L_2 C_5 R_S + C_1 L_4 C_5 R_S + C_3 L_4 C_5 R_S + L_2 C_3 L_4 \frac{1}{R_L} \right) \\ + \omega^2 \left(L_2 C_3 + L_2 C_5 + L_4 C_5 + C_1 L_2 \frac{R_S}{R_L} + C_1 L_4 \frac{R_S}{R_L} + C_3 L_4 \frac{R_S}{R_L} \right) \\ + \omega \left(C_1 R_S + \frac{L_2}{R_L} + C_3 R_S + \frac{L_4}{R_L} + C_5 R_S \right) + \left(\frac{R_S}{R_L} + 1 \right) \end{array}} \tag{10.1}$$

The corresponding normalized Butterworth transfer function using the product formula of Eq. (2.1) is

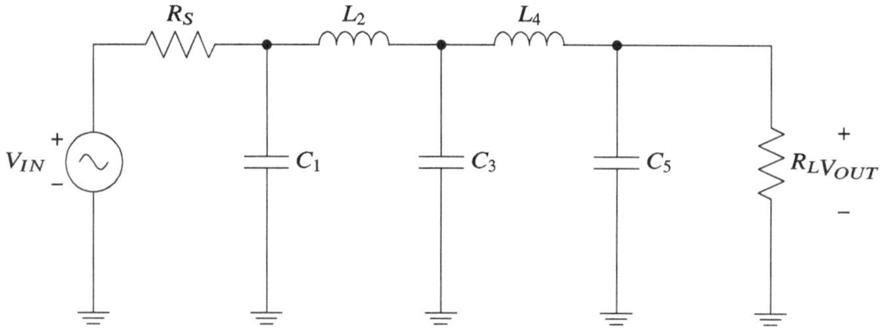

Fig. 10.1 Fifth-order shunt-first doubly-terminated low-pass Butterworth filter

$$\frac{V_{OUT}}{V_{IN}} = \frac{1}{\begin{aligned}&s^5\left(\left(\frac{\cos(0)}{\sin\left(\frac{1}{10}\pi\right)}\right)\left(\frac{\cos\left(\frac{1}{10}\pi\right)}{\sin\left(\frac{2}{10}\pi\right)}\right)\left(\frac{\cos\left(\frac{2}{10}\pi\right)}{\sin\left(\frac{3}{10}\pi\right)}\right)\left(\frac{\cos\left(\frac{3}{10}\pi\right)}{\sin\left(\frac{4}{10}\pi\right)}\right)\left(\frac{\cos\left(\frac{4}{10}\pi\right)}{\sin\left(\frac{5}{10}\pi\right)}\right)\right)\\&+s^4\left(\left(\frac{\cos(0)}{\sin\left(\frac{1}{10}\pi\right)}\right)\left(\frac{\cos\left(\frac{1}{10}\pi\right)}{\sin\left(\frac{2}{10}\pi\right)}\right)\left(\frac{\cos\left(\frac{2}{10}\pi\right)}{\sin\left(\frac{3}{10}\pi\right)}\right)\left(\frac{\cos\left(\frac{3}{10}\pi\right)}{\sin\left(\frac{4}{10}\pi\right)}\right)\right)\\&+s^3\left(\left(\frac{\cos(0)}{\sin\left(\frac{1}{10}\pi\right)}\right)\left(\frac{\cos\left(\frac{1}{10}\pi\right)}{\sin\left(\frac{2}{10}\pi\right)}\right)\left(\frac{\cos\left(\frac{2}{10}\pi\right)}{\sin\left(\frac{3}{10}\pi\right)}\right)\right)\\&+s^2\left(\left(\frac{\cos(0)}{\sin\left(\frac{1}{10}\pi\right)}\right)\left(\frac{\cos\left(\frac{1}{10}\pi\right)}{\sin\left(\frac{2}{10}\pi\right)}\right)\right)\\&+s\left(\left(\frac{\cos(0)}{\sin\left(\frac{1}{10}\pi\right)}\right)\right)+1\end{aligned}} \tag{10.2}$$

The fifth-degree term of the realized denominator,

$$\omega^5 C_1 L_2 C_3 L_4 C_5 R_S,$$

is associated with the fifth-degree Butterworth term

$$s^5\left(\left(\frac{\cos(0)}{\sin\left(\frac{1}{10}\pi\right)}\right)\left(\frac{\cos\left(\frac{1}{10}\pi\right)}{\sin\left(\frac{2}{10}\pi\right)}\right)\left(\frac{\cos\left(\frac{2}{10}\pi\right)}{\sin\left(\frac{3}{10}\pi\right)}\right)\left(\frac{\cos\left(\frac{3}{10}\pi\right)}{\sin\left(\frac{4}{10}\pi\right)}\right)\left(\frac{\cos\left(\frac{4}{10}\pi\right)}{\sin\left(\frac{5}{10}\pi\right)}\right)\right).$$

The fourth-degree term of the realized denominator,

$$\omega^4\left(C_1 L_2 C_3 L_4 \frac{R_S}{R_L} + L_2 C_3 L_4 C_5\right),$$

is associated with the fourth-degree Butterworth term

$$s^4\left(\left(\frac{\cos(0)}{\sin\left(\frac{1}{10}\pi\right)}\right)\left(\frac{\cos\left(\frac{1}{10}\pi\right)}{\sin\left(\frac{2}{10}\pi\right)}\right)\left(\frac{\cos\left(\frac{2}{10}\pi\right)}{\sin\left(\frac{3}{10}\pi\right)}\right)\left(\frac{\cos\left(\frac{3}{10}\pi\right)}{\sin\left(\frac{4}{10}\pi\right)}\right)\right).$$

The third-degree term of the realized denominator,

$$\omega^3\left(C_1 L_2 C_3 R_S + C_1 L_2 C_5 R_S + C_1 L_4 C_5 R_S + C_3 L_4 C_5 R_S + L_2 C_3 L_4 \frac{1}{R_L}\right),$$

is associated with the third-degree Butterworth term

$$s^3\left(\left(\frac{\cos(0)}{\sin\left(\frac{1}{10}\pi\right)}\right)\left(\frac{\cos(\frac{1}{10}\pi)}{\sin\left(\frac{2}{10}\pi\right)}\right)\left(\frac{\cos\left(\frac{2}{10}\pi\right)}{\sin\left(\frac{3}{10}\pi\right)}\right)\right).$$

The second-degree term of the realized denominator,

$$\omega^2\left(L_2C_3+L_2C_5+L_4C_5+C_1L_2\frac{R_S}{R_L}+C_1L_4\frac{R_S}{R_L}+C_3L_4\frac{R_S}{R_L}\right),$$

is associated with the second-degree Butterworth term

$$s^2\left(\left(\frac{\cos(0)}{\sin\left(\frac{1}{10}\pi\right)}\right)\left(\frac{\cos(\frac{1}{10}\pi)}{\sin\left(\frac{2}{10}\pi\right)}\right)\right).$$

The first-degree term of the realized denominator,

$$\omega\left(C_1R_S+\frac{L_2}{R_L}+C_3R_S+\frac{L_4}{R_L}+C_5R_S\right),$$

is associated with the first-degree Butterworth term

$$s\left(\left(\frac{\cos(0)}{\sin\left(\frac{1}{10}\pi\right)}\right)\right).$$

Applying the transformation rule yields

$$\begin{aligned}
AC_{aspect} &= \left(DC_{aspect}\right)\left(Butterworth_{coef}\right)\\
\omega^5C_1L_2C_3L_4C_5R_S &= \left(\frac{R_S}{R_L}+1\right)\\
&\times\left(\frac{\cos(0)}{\sin\left(\frac{1}{10}\pi\right)}\right)\left(\frac{\cos(\frac{1}{10}\pi)}{\sin\left(\frac{2}{10}\pi\right)}\right)\\
&\times\left(\frac{\cos\left(\frac{2}{10}\pi\right)}{\sin\left(\frac{3}{10}\pi\right)}\right)\left(\frac{\cos(\frac{3}{10}\pi)}{\sin\left(\frac{4}{10}\pi\right)}\right)\left(\frac{\cos(\frac{4}{10}\pi)}{\sin\left(\frac{5}{10}\pi\right)}\right)\\
\omega^4\left(C_1L_2C_3L_4\frac{R_S}{R_L}+L_2C_3L_4C_5\right) &= \left(\frac{R_S}{R_L}+1\right)\\
&\times\left(\frac{\cos(0)}{\sin\left(\frac{1}{10}\pi\right)}\right)\left(\frac{\cos(\frac{1}{10}\pi)}{\sin\left(\frac{2}{10}\pi\right)}\right)\\
&\times\left(\frac{\cos\left(\frac{2}{10}\pi\right)}{\sin\left(\frac{3}{10}\pi\right)}\right)\left(\frac{\cos(\frac{3}{10}\pi)}{\sin\left(\frac{4}{10}\pi\right)}\right)\\
\omega^3\,(C_1L_2C_3R_S &+ C_1L_2C_5R_S+C_1L_4C_5R_S
\end{aligned}$$

$$
\begin{aligned}
&+ C_3L_4C_5R_S + L_2C_3L_4\frac{1}{R_L}\Bigg) = \left(\frac{R_S}{R_L}+1\right)\\
&\times \left(\frac{\cos(0)}{\sin\left(\frac{1}{10}\pi\right)}\right)\left(\frac{\cos(\frac{1}{10}\pi)}{\sin\left(\frac{2}{10}\pi\right)}\right)\\
&\times \left(\frac{\cos\left(\frac{2}{10}\pi\right)}{\sin\left(\frac{3}{10}\pi\right)}\right)\\
\omega^2 \Bigg(&L_2C_3 + L_2C_5 + L_4C_5 + C_1L_2\frac{R_S}{R_L}\\
&+ C_1L_4\frac{R_S}{R_L} + C_3L_4\frac{R_S}{R_L}\Bigg) = \left(\frac{R_S}{R_L}+1\right)\\
&\times \left(\frac{\cos(0)}{\sin\left(\frac{1}{10}\pi\right)}\right)\left(\frac{\cos(\frac{1}{10}\pi)}{\sin\left(\frac{2}{10}\pi\right)}\right)\\
\omega &\left(C_1R_S + \frac{L_2}{R_L} + C_3R_S + \frac{L_4}{R_L} + C_5R_S\right) = \left(\frac{R_S}{R_L}+1\right)\\
&\times \left(\frac{\cos(0)}{\sin\left(\frac{1}{10}\pi\right)}\right)
\end{aligned}
\tag{10.3}
$$

The system of Eqs. (10.3) simplifies to

$$
\omega^5C_1L_2C_3L_4C_5R_S = \left(\frac{R_S}{R_L}+1\right), \tag{10.4}
$$

$$
\omega^4\left(C_1L_2C_3L_4\frac{R_S}{R_L} + L_2C_3L_4C_5\right) = \left(\csc(\frac{1}{10}\pi)\right)\left(\frac{R_S}{R_L}+1\right) = (2\Phi)\left(\frac{R_S}{R_L}+1\right), \tag{10.5}
$$

$$
\begin{aligned}
\omega^3 &\left(C_1L_2C_3R_S + C_1L_2C_5R_S + C_1L_4C_5R_S + C_3L_4C_5R_S + L_2C_3L_4\frac{1}{R_L}\right)\\
&= \left(\frac{\csc(\frac{1}{10}\pi)^2}{2}\right)\left(\frac{R_S}{R_L}+1\right) = (2\Phi^2)\left(\frac{R_S}{R_L}+1\right),
\end{aligned}
\tag{10.6}
$$

$$
\begin{aligned}
\omega^2 &\left(L_2C_3 + L_2C_5 + L_4C_5 + C_1L_2\frac{R_S}{R_L} + C_1L_4\frac{R_S}{R_L} + C_3L_4\frac{R_S}{R_L}\right)\\
&= \left(\frac{\csc(\frac{1}{10}\pi)^2}{2}\right)\left(\frac{R_S}{R_L}+1\right) = (2\Phi^2)\left(\frac{R_S}{R_L}+1\right),
\end{aligned}
\tag{10.7}
$$

and

$$\omega\left(C_1R_S + \frac{L_2}{R_L} + C_3R_S + \frac{L_4}{R_L} + C_5R_S\right) = \left(\csc(\frac{1}{10}\pi)\right)\left(\frac{R_S}{R_L}+1\right) = (2\Phi)\left(\frac{R_S}{R_L}+1\right). \tag{10.8}$$

The fifth-degree term and the constant term are equal. The fourth- and first-degree terms are equal. The third- and second-degree terms are equal. This is the denominator-level manifestation of fifth-order odd symmetry.

Since there are five equations and five unknown reactive element values, the fifth-order system can produce multiple candidate solution tuples. In practice, the real-valued tuples are retained as exact passive realizations.

This behavior should be distinguished from a failure of the coefficient-matching method. The equations still produce real-valued passive solution tuples. What changes at fifth order is that the candidate solution set becomes more complicated at the same point where the secondary correlated factor family first appears in the product-factor table.

10.2.1 Design Example Fifth-Order Low-Pass Shunt-First

The design requirements are:

- $R_S = 510\,\text{k}\Omega$
- $R_L = 511\,\text{k}\Omega$
- $f_c = 343\,\text{kHz}$
- $\omega_c = 2.155132 \times 10^6\,\text{rad/s}$.

The gain at DC, or throughout the passband as established by the terminations, is

$$20\log_{10}\left(\frac{511\,\text{k}}{510\,\text{k} + 511\,\text{k}}\right) = -6.012096\,\text{dB}.$$

Therefore the gain at cutoff is

$$20\log_{10}\left(\frac{1}{\sqrt{2}} * \frac{511\,\text{k}}{511\,\text{k} + 510\,\text{k}}\right) = -9.022395\,\text{dB},$$

or

$$-6.012096\,\text{dB} - 3.010299\,\text{dB} = -9.022395\,\text{dB}.$$

A MATLAB implementation of the system of equations in Eq. 10.3 may be written as

```
format longEng

Rs = 510e3;
Rl = 511e3;
```

```
f = 343e3;
w = 2*pi*f;

syms C1 L2 C3 L4 C5

E0 = ((Rs/Rl)+1)

E1 = w^1*(C1*Rs+L2/Rl+C3*Rs+L4/Rl+C5*Rs) ==
E0*(cos(0)/sin((1*pi)/10))

E2 = w^2*(C1*L2*Rs/Rl+C1*L4*Rs/Rl+C3*L2+C3*L4*Rs/Rl+C5*L2+C5*L4) ==
E0*(cos(0)/sin((1*pi)/10))*
(cos((1*pi)/10)/sin((2*pi)/10))

E3 = w^3*(C1*L2*C3*Rs+C1*L2*C5*Rs+C1*L4*C5*Rs+C3*L4*C5*Rs+(L2*C3*L4)/Rl) ==
E0*(cos(0)/sin((1*pi)/10))*
(cos((1*pi)/10)/sin((2*pi)/10))*
(cos((2*pi)/10)/sin((3*pi)/10))

E4 = w^4*(C1*L2*C3*L4*Rs/Rl+L2*C3*L4*C5) ==
E0*(cos(0)/sin((1*pi)/10))*
(cos((1*pi)/10)/sin((2*pi)/10))*
(cos((2*pi)/10)/sin((3*pi)/10))*
(cos((3*pi)/10)/sin((4*pi)/10))

E5 = w^5*(C1*L2*C3*L4*C5*Rs) ==
E0*(cos(0)/sin((1*pi)/10))*
(cos((1*pi)/10)/sin((2*pi)/10))*
(cos((2*pi)/10)/sin((3*pi)/10))*
(cos((3*pi)/10)/sin((4*pi)/10))*
(cos((4*pi)/10)/sin((5*pi)/10))


result = solve(E1,E2,E3,E4,E5);

C1 = vpa(result.C1)
L2 = vpa(result.L2)
C3 = vpa(result.C3)
L4 = vpa(result.L4)
C5 = vpa(result.C5)
```

The complete candidate solution tuple output for this design example is provided in Appendix B, Sect. B.13.

In this example, the solver returns sixteen candidate tuples: four fully real tuples and six complex-conjugate solution pairs. The four fully real tuples are retained as valid passive realizations.

To provide a concrete component set for verification, one of the fully real tuples is selected as a representative passive realization. This selection is not unique; any of the fully real tuples may be used to satisfy the design requirements. The selected tuple is (Fig. 10.2)

$$C_1 = 449.7873\,\text{fF}, \quad L_2 = 326.2272\,\text{mH}, \quad C_3 = 1.754678\,\text{pF},$$
$$L_4 = 437.3195\,\text{mH}, \quad C_5 = 748.4128\,\text{fF}.$$

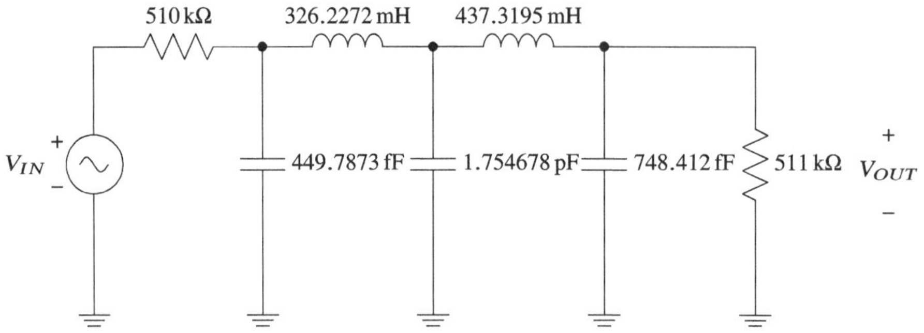

Fig. 10.2 Representative fifth-order shunt-first low-pass solution

10.3 Low-Pass Series-First

The transfer function for the fifth-order series-first doubly-terminated low-pass Butterworth filter in Fig. 10.3 is

$$\frac{V_{OUT}}{V_{IN}} = \frac{1}{\begin{aligned}&\omega^5 L_1 C_2 L_3 C_4 L_5 \frac{1}{R_L} \\ &+ \omega^4 \left(L_1 C_2 L_3 C_4 + C_2 L_3 C_4 L_5 \frac{R_S}{R_L} \right) \\ &+ \omega^3 \left(\frac{L_1 C_2 L_3}{R_L} + \frac{L_1 C_2 L_5}{R_L} + \frac{L_1 C_4 L_5}{R_L} + \frac{L_3 C_4 L_5}{R_L} + C_2 L_3 C_4 R_S \right) \\ &+ \omega^2 \left(L_1 C_2 + L_1 C_4 + L_3 C_4 + C_2 L_3 \frac{R_S}{R_L} + C_2 L_5 \frac{R_S}{R_L} + C_4 L_5 \frac{R_S}{R_L} \right) \\ &+ \omega \left(\frac{L_1}{R_L} + C_2 R_S + \frac{L_3}{R_L} + C_4 R_S + \frac{L_5}{R_L} \right) + \left(\frac{R_S}{R_L} + 1 \right)\end{aligned}} \tag{10.9}$$

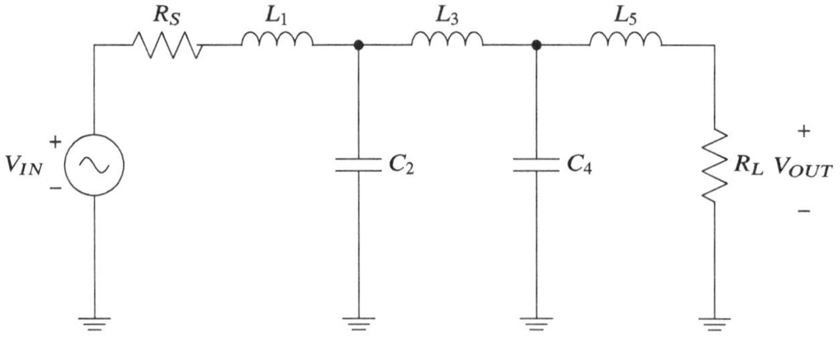

Fig. 10.3 Fifth-order series-first doubly-terminated low-pass Butterworth filter

The corresponding normalized Butterworth transfer function using the product formula of Eq. (2.1) is

$$\frac{V_{OUT}}{V_{IN}} = \frac{1}{\begin{aligned} & s^5\left(\left(\frac{\cos(0)}{\sin\left(\frac{1}{10}\pi\right)}\right)\left(\frac{\cos(\frac{1}{10}\pi)}{\sin\left(\frac{2}{10}\pi\right)}\right)\left(\frac{\cos\left(\frac{2}{10}\pi\right)}{\sin\left(\frac{3}{10}\pi\right)}\right)\left(\frac{\cos(\frac{3}{10}\pi)}{\sin\left(\frac{4}{10}\pi\right)}\right)\left(\frac{\cos(\frac{4}{10}\pi)}{\sin\left(\frac{5}{10}\pi\right)}\right)\right) \\ & + s^4\left(\left(\frac{\cos(0)}{\sin\left(\frac{1}{10}\pi\right)}\right)\left(\frac{\cos(\frac{1}{10}\pi)}{\sin\left(\frac{2}{10}\pi\right)}\right)\left(\frac{\cos\left(\frac{2}{10}\pi\right)}{\sin\left(\frac{3}{10}\pi\right)}\right)\left(\frac{\cos(\frac{3}{10}\pi)}{\sin\left(\frac{4}{10}\pi\right)}\right)\right) \\ & + s^3\left(\left(\frac{\cos(0)}{\sin\left(\frac{1}{10}\pi\right)}\right)\left(\frac{\cos(\frac{1}{10}\pi)}{\sin\left(\frac{2}{10}\pi\right)}\right)\left(\frac{\cos\left(\frac{2}{10}\pi\right)}{\sin\left(\frac{3}{10}\pi\right)}\right)\right) \\ & + s^2\left(\left(\frac{\cos(0)}{\sin\left(\frac{1}{10}\pi\right)}\right)\left(\frac{\cos(\frac{1}{10}\pi)}{\sin\left(\frac{2}{10}\pi\right)}\right)\right) \\ & + s\left(\left(\frac{\cos(0)}{\sin\left(\frac{1}{10}\pi\right)}\right)\right) + 1 \end{aligned}} \tag{10.10}$$

As expected from the companion ladder topology, the same fifth-order Butterworth coefficient sequence is used; the distinction is the series-first form of the realized denominator.

The fifth-degree term of the realized denominator,

$$\omega^5 L_1 C_2 L_3 C_4 L_5 \frac{1}{R_L},$$

is associated with the fifth-degree Butterworth term

$$s^5\left(\left(\frac{\cos(0)}{\sin\left(\frac{1}{10}\pi\right)}\right)\left(\frac{\cos(\frac{1}{10}\pi)}{\sin\left(\frac{2}{10}\pi\right)}\right)\left(\frac{\cos\left(\frac{2}{10}\pi\right)}{\sin\left(\frac{3}{10}\pi\right)}\right)\left(\frac{\cos(\frac{3}{10}\pi)}{\sin\left(\frac{4}{10}\pi\right)}\right)\left(\frac{\cos(\frac{4}{10}\pi)}{\sin\left(\frac{5}{10}\pi\right)}\right)\right).$$

The fourth-degree term of the realized denominator,

$$\omega^4\left(L_1 C_2 L_3 C_4 + C_2 L_3 C_4 L_5 \frac{R_S}{R_L}\right),$$

is associated with the fourth-degree Butterworth term

$$s^4\left(\left(\frac{\cos(0)}{\sin\left(\frac{1}{10}\pi\right)}\right)\left(\frac{\cos(\frac{1}{10}\pi)}{\sin\left(\frac{2}{10}\pi\right)}\right)\left(\frac{\cos\left(\frac{2}{10}\pi\right)}{\sin\left(\frac{3}{10}\pi\right)}\right)\left(\frac{\cos(\frac{3}{10}\pi)}{\sin\left(\frac{4}{10}\pi\right)}\right)\right).$$

The third-degree term of the realized denominator,

$$\omega^3\left(\frac{L_1C_2L_3}{R_L}+\frac{L_1C_2L_5}{R_L}+\frac{L_1C_4L_5}{R_L}+\frac{L_3C_4L_5}{R_L}+C_2L_3C_4R_S\right),$$

is associated with the third-degree Butterworth term

$$s^3\left(\left(\frac{\cos(0)}{\sin\left(\frac{1}{10}\pi\right)}\right)\left(\frac{\cos(\frac{1}{10}\pi)}{\sin\left(\frac{2}{10}\pi\right)}\right)\left(\frac{\cos\left(\frac{2}{10}\pi\right)}{\sin\left(\frac{3}{10}\pi\right)}\right)\right).$$

The second-degree term of the realized denominator,

$$\omega^2\left(L_1C_2+L_1C_4+L_3C_4+C_2L_3\frac{R_S}{R_L}+C_2L_5\frac{R_S}{R_L}+C_4L_5\frac{R_S}{R_L}\right),$$

is associated with the second-degree Butterworth term

$$s^2\left(\left(\frac{\cos(0)}{\sin\left(\frac{1}{10}\pi\right)}\right)\left(\frac{\cos(\frac{1}{10}\pi)}{\sin\left(\frac{2}{10}\pi\right)}\right)\right).$$

The first-degree term of the realized denominator,

$$\omega\left(\frac{L_1}{R_L}+C_2R_S+\frac{L_3}{R_L}+C_4R_S+\frac{L_5}{R_L}\right),$$

is associated with the first-degree Butterworth term

$$s\left(\left(\frac{\cos(0)}{\sin\left(\frac{1}{10}\pi\right)}\right)\right).$$

Applying the transformation rule yields

$$
\begin{aligned}
&AC_{aspect} = \left(DC_{aspect}\right)\left(Butterworth_{coef}\right)\\
&\omega^5 L_1 C_2 L_3 C_4 L_5 \frac{1}{R_L} = \left(\frac{R_S}{R_L}+1\right)\\
&\quad \left(\left(\frac{\cos(0)}{\sin\left(\frac{1}{10}\pi\right)}\right)\left(\frac{\cos(\frac{1}{10}\pi)}{\sin\left(\frac{2}{10}\pi\right)}\right)\left(\frac{\cos\left(\frac{2}{10}\pi\right)}{\sin\left(\frac{3}{10}\pi\right)}\right)\left(\frac{\cos(\frac{3}{10}\pi)}{\sin\left(\frac{4}{10}\pi\right)}\right)\left(\frac{\cos(\frac{4}{10}\pi)}{\sin\left(\frac{5}{10}\pi\right)}\right)\right)\\
&\omega^4\left(L_1C_2L_3C_4 + C_2L_3C_4L_5\frac{R_S}{R_L}\right)\\
&\quad = \left(\frac{R_S}{R_L}+1\right)\left(\left(\frac{\cos(0)}{\sin\left(\frac{1}{10}\pi\right)}\right)\left(\frac{\cos(\frac{1}{10}\pi)}{\sin\left(\frac{2}{10}\pi\right)}\right)\left(\frac{\cos\left(\frac{2}{10}\pi\right)}{\sin\left(\frac{3}{10}\pi\right)}\right)\left(\frac{\cos(\frac{3}{10}\pi)}{\sin\left(\frac{4}{10}\pi\right)}\right)\right)\\
&\omega^3\left(\frac{L_1C_2L_3}{R_L}+\frac{L_1C_2L_5}{R_L}+\frac{L_1C_4L_5}{R_L}+\frac{L_3C_4L_5}{R_L}+C_2L_3C_4R_S\right)\\
&\quad = \left(\frac{R_S}{R_L}+1\right)\left(\left(\frac{\cos(0)}{\sin\left(\frac{1}{10}\pi\right)}\right)\left(\frac{\cos(\frac{1}{10}\pi)}{\sin\left(\frac{2}{10}\pi\right)}\right)\left(\frac{\cos\left(\frac{2}{10}\pi\right)}{\sin\left(\frac{3}{10}\pi\right)}\right)\right)\\
&\omega^2\left(L_1C_2 + L_1C_4 + L_3C_4 + C_2L_3\frac{R_S}{R_L} + C_2L_5\frac{R_S}{R_L} + C_4L_5\frac{R_S}{R_L}\right)\\
&\quad = \left(\frac{R_S}{R_L}+1\right)\left(\left(\frac{\cos(0)}{\sin\left(\frac{1}{10}\pi\right)}\right)\left(\frac{\cos(\frac{1}{10}\pi)}{\sin\left(\frac{2}{10}\pi\right)}\right)\right)\\
&\omega\left(\frac{L_1}{R_L}+C_2R_S+\frac{L_3}{R_L}+C_4R_S+\frac{L_5}{R_L}\right) = \left(\frac{R_S}{R_L}+1\right)\left(\left(\frac{\cos(0)}{\sin\left(\frac{1}{10}\pi\right)}\right)\right)
\end{aligned}
\tag{10.11}
$$

The system of Eqs. (10.11) simplifies to

$$
\omega^5 L_1C_2L_3C_4L_5\frac{1}{R_L} = \left(\frac{R_S}{R_L}+1\right), \tag{10.12}
$$

$$
\omega^4\left(L_1C_2L_3C_4 + C_2L_3C_4L_5\frac{R_S}{R_L}\right) = \left(\csc(\frac{1}{10}\pi)\right)\left(\frac{R_S}{R_L}+1\right) = (2\Phi)\left(\frac{R_S}{R_L}+1\right), \tag{10.13}
$$

$$
\begin{aligned}
&\omega^3\left(\frac{L_1C_2L_3}{R_L}+\frac{L_1C_2L_5}{R_L}+\frac{L_1C_4L_5}{R_L}+\frac{L_3C_4L_5}{R_L}+C_2L_3C_4R_S\right)\\
&\quad = \left(\frac{\csc(\frac{1}{10}\pi)^2}{2}\right)\left(\frac{R_S}{R_L}+1\right) = (2\Phi^2)\left(\frac{R_S}{R_L}+1\right),
\end{aligned}
\tag{10.14}
$$

$$\omega^2\left(L_1C_2+L_1C_4+L_3C_4+C_2L_3\frac{R_S}{R_L}+C_2L_5\frac{R_S}{R_L}+C_4L_5\frac{R_S}{R_L}\right)$$
$$=\left(\frac{\csc(\frac{1}{10}\pi)^2}{2}\right)\left(\frac{R_S}{R_L}+1\right)=(2\Phi^2)\left(\frac{R_S}{R_L}+1\right), \tag{10.15}$$

and

$$\omega\left(\frac{L_1}{R_L}+C_2R_S+\frac{L_3}{R_L}+C_4R_S+\frac{L_5}{R_L}\right)=\left(\csc(\frac{1}{10}\pi)\right)\left(\frac{R_S}{R_L}+1\right)=(2\Phi)\left(\frac{R_S}{R_L}+1\right). \tag{10.16}$$

As in the shunt-first case, the fifth-degree term and the constant term are equal. The fourth- and first-degree terms are equal. The third- and second-degree terms are equal. This is the denominator-level manifestation of fifth-order odd symmetry.

Since there are five equations and five unknown reactive element values, the fifth-order series-first system can produce multiple candidate solution tuples. In practice, the real-valued tuples are retained as exact passive realizations.

As with the shunt-first topology, this increased candidate-solution structure is treated as an observed fifth-order feature rather than as a proven limitation of the synthesis method.

10.3.1 Design Example Fifth-Order Low-Pass Series-First

The design requirements are:

- $R_S = 2.002\,\text{k}\Omega$
- $R_L = 2.001\,\text{k}\Omega$
- $f_c = 190.9859\,\text{Hz}$
- $\omega_c = 1.2 \times 10^3\,\text{rad/s}$.

The gain at DC, or throughout the passband as established by the terminations, is

$$20\log_{10}\left(\frac{2.001\,\text{k}\Omega}{2.002\,\text{k}\Omega+2.001\,\text{k}\Omega}\right)=-6.022770\,\text{dB},$$

Therefore the gain at cutoff is

$$20\log_{10}\left(\frac{1}{\sqrt{2}}*\frac{2.001\,\text{k}\Omega}{2.001\,\text{k}\Omega+2.002\,\text{k}}\right)=-9.033069\,\text{dB},$$

or

$$-6.022770\,\text{dB}-3.010299\,\text{dB}=-9.033069\,\text{dB}.$$

A MATLAB implementation of the system of equations in Eq. (10.11) may be written as

```
format longENG

Rs = 2.002e3
Rl = 2.001e3
f = 190.9859
w = 1200

syms L1 C2 L3 C4 L5

E0 = ((Rs/Rl)+1)

E1 = w^1*(L1/Rl + C2*Rs + L3/Rl + C4*Rs + L5/Rl) ==
   E0*(cos(0)/sin((1*pi)/10))

E2 = w^2*(L1*C2 + L1*C4 + L3*C2*Rs/Rl + L3*C4 + L5*C2*Rs/Rl + L5*C4*Rs/Rl) ==
   E0*(cos(0)/sin((1*pi)/10))*
     (cos((1*pi)/10)/sin((2*pi)/10))

E3 = w^3*(L1*C2*L3/Rl + L1*C2*L5/Rl + L1*C4*L5/Rl + C2*L3*C4*Rs + L3*C4*L5/Rl) ==
   E0*(cos(0)/sin((1*pi)/10))*
     (cos((1*pi)/10)/sin((2*pi)/10))*
     (cos((2*pi)/10)/sin((3*pi)/10))

E4 = w^4*(L1*C2*L3*C4+C2*L3*C4*L5*Rs/Rl) ==
   E0*(cos(0)/sin((1*pi)/10))*
     (cos((1*pi)/10)/sin((2*pi)/10))*
     (cos((2*pi)/10)/sin((3*pi)/10))*
     (cos((3*pi)/10)/sin((4*pi)/10))

E5 = w^5*(L1*C2*L3*C4*L5)/Rl ==
   E0*(cos(0)/sin((1*pi)/10))*
     (cos((1*pi)/10)/sin((2*pi)/10))*
     (cos((2*pi)/10)/sin((3*pi)/10))*
     (cos((3*pi)/10)/sin((4*pi)/10))*
     (cos((4*pi)/10)/sin((5*pi)/10))

digits(10)
result = vpasolve(E1,E2,E3,E4,E5)

L1 = vpa(result.L1)
C2 = vpa(result.C2)
L3 = vpa(result.L3)
C4 = vpa(result.C4)
L5 = vpa(result.L5)

M_L1 = abs(L1)
M_C2 = abs(C2)
M_L3 = abs(L3)
M_C4 = abs(C4)
M_L5 = abs(L5)
```

The complete candidate solution tuple output for this design example is provided in Appendix B, Sect. B.14.

In this example, the solver returns sixteen candidate tuples: four fully real tuples and six complex-conjugate solution pairs. Each complex-conjugate pair also produces a magnitude tuple. These magnitude tuples are less accurate than the fully real tuples and are not used as the representative passive realization. The four fully real tuples are retained as valid passive realizations.

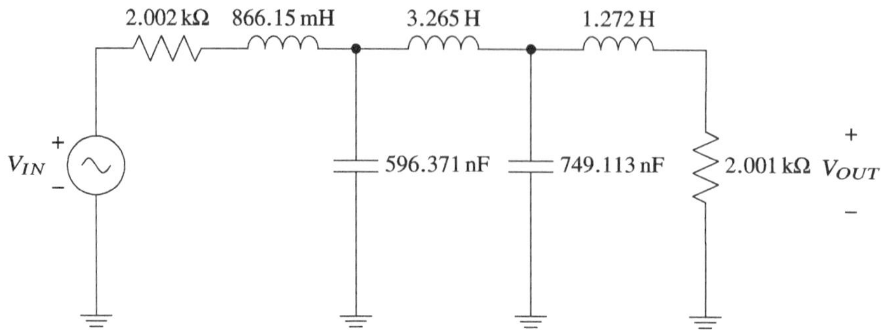

Fig. 10.4 Representative fifth-order series-first low-pass solution

To provide a concrete component set for verification, one of the fully real tuples is selected as a representative passive realization. This selection is not unique; any of the fully real tuples may be used to satisfy the design requirements. The selected tuple is (Fig. 10.4)

$$L_1 = 866.150699\,\text{mH}, \quad C_2 = 596.371096\,\text{nF}, \quad L_3 = 3.26528208\,\text{H},$$

$$C_4 = 749.113095\,\text{nF}, \quad L_5 = 1.272746708\,\text{H}.$$

10.4 Time Constants

Equations (6.8) and (6.17) define τ_C and τ_L, which are used below.

For the fifth-order shunt-first low-pass filter, the first-degree portion of the system is

$$\begin{aligned}\omega\left(C_1R_S + \frac{L_2}{R_L} + C_3R_S + \frac{L_4}{R_L} + C_5R_S\right) &= (2\Phi)\left(\frac{R_S}{R_L}+1\right)\\ &= \csc\left(\frac{\pi}{10}\right)\left(\frac{R_S}{R_L}+1\right),\\ \omega\left(C_1\frac{R_SR_L}{R_S+R_L} + \frac{L_2}{R_S+R_L} + C_3\frac{R_SR_L}{R_S+R_L}\right.&\\ \left. + \frac{L_4}{R_S+R_L} + C_5\frac{R_SR_L}{R_S+R_L}\right) &= 2\Phi = \csc\left(\frac{\pi}{10}\right).\end{aligned} \tag{10.17}$$

For the fifth-order series-first low-pass filter, the corresponding first-degree portion is

$$\begin{aligned}
\omega\left(\frac{L_1}{R_L}+C_2R_S+\frac{L_3}{R_L}+C_4R_S+\frac{L_5}{R_L}\right)&=(2\Phi)\left(\frac{R_S}{R_L}+1\right)\\
&=\csc\left(\frac{\pi}{10}\right)\left(\frac{R_S}{R_L}+1\right),\\
\omega\left(\frac{L_1}{R_S+R_L}+C_2\frac{R_SR_L}{R_S+R_L}+\frac{L_3}{R_S+R_L}\right.&\\
\left.+C_4\frac{R_SR_L}{R_S+R_L}+\frac{L_5}{R_S+R_L}\right)&=2\Phi=\csc\left(\frac{\pi}{10}\right).
\end{aligned} \tag{10.18}$$

Thus, both fifth-order topologies resolve to

$$\omega\left(\tau_{C_1}+\tau_{L_2}+\tau_{C_3}+\tau_{L_4}+\tau_{C_5}\right)=\omega\left(\tau_{L_1}+\tau_{C_2}+\tau_{L_3}+\tau_{C_4}+\tau_{L_5}\right)=2\Phi=\csc\left(\frac{\pi}{10}\right).$$

If $R_S=R_L=1$, then

$$\begin{aligned}
\omega\tau_{C_1}&=\omega\tau_{L_1}=\sin\left(\frac{1}{10}\pi\right),\\
\omega\tau_{L_2}&=\omega\tau_{C_2}=\sin\left(\frac{3}{10}\pi\right),\\
\omega\tau_{C_3}&=\omega\tau_{L_3}=\sin\left(\frac{5}{10}\pi\right),\\
\omega\tau_{L_4}&=\omega\tau_{C_4}=\sin\left(\frac{7}{10}\pi\right),\\
\omega\tau_{C_5}&=\omega\tau_{L_5}=\sin\left(\frac{9}{10}\pi\right).
\end{aligned}$$

Therefore,

$$\begin{aligned}
&\sin\left(\frac{1}{10}\pi\right)+\sin\left(\frac{3}{10}\pi\right)+\sin\left(\frac{5}{10}\pi\right)+\sin\left(\frac{7}{10}\pi\right)+\sin\left(\frac{9}{10}\pi\right)\\
&\quad=\frac{1}{\sin\left(\frac{1}{10}\pi\right)}=\csc\left(\frac{1}{10}\pi\right)=2\Phi.
\end{aligned} \tag{10.19}$$

If $R_S\neq R_L$, then

$$\omega\tau_{C_1} \neq \omega\tau_{L_1} \neq \sin\left(\frac{1}{10}\pi\right),$$
$$\omega\tau_{L_2} \neq \omega\tau_{C_2} \neq \sin\left(\frac{3}{10}\pi\right),$$
$$\omega\tau_{C_3} \neq \omega\tau_{L_3} \neq \sin\left(\frac{5}{10}\pi\right),$$
$$\omega\tau_{L_4} \neq \omega\tau_{C_4} \neq \sin\left(\frac{7}{10}\pi\right),$$
$$\omega\tau_{C_5} \neq \omega\tau_{L_5} \neq \sin\left(\frac{9}{10}\pi\right).$$

For $R_S \neq R_L$, the individual time constants no longer equal the corresponding sine terms, but the summed identity of Eq. (10.19) remains the reference pattern for the fifth-order case. This is the fifth-order instance of the finite trigonometric time-constant structure that is later collected in Chap. 12.

10.5 High-Pass Shunt-First

A fifth-order shunt-first high-pass filter is obtained by first synthesizing the corresponding fifth-order shunt-first low-pass ladder and then converting each reactive element using Eqs. (6.19) and (6.20). Therefore, the high-pass example is not a separate coefficient-matching derivation. It is a transformation example that uses the shunt-first low-pass system of Eq. (10.3) to obtain a low-pass precursor, which is then converted into the final high-pass ladder. Low-pass shunt capacitors become high-pass shunt inductors, and low-pass series inductors become high-pass series capacitors.

10.5.1 Design Example Fifth-Order High-Pass Shunt-First

The design requirements are:

- $R_S = 7\,\text{k}\Omega$
- $R_L = 2\,\text{k}\Omega$
- $f_c = 70.02817\,\text{kHz}$
- $\omega_c = 440 \times 10^3\,\text{rad/s}$.

The passband gain established by the terminations is

$$20\log_{10}\left(\frac{2\,\text{k}\Omega}{7\,\text{k}\Omega + 2\,\text{k}\Omega}\right) = -13.06425\,\text{dB}.$$

Therefore the gain at the high-pass cut-on frequency is

$$20\log_{10}\left(\frac{1}{\sqrt{2}} * \frac{2\,\text{k}\Omega}{7\,\text{k}\Omega + 2\,\text{k}\Omega}\right) = -16.07455\,\text{dB},$$

or

$$-\,13.06425\,\text{dB} - 3.010299\,\text{dB} = -16.07455\,\text{dB}.$$

To obtain the low-pass precursor for the high-pass realization, the shunt-first low-pass system of Eq. (10.3) is solved using the high-pass design requirements. A MATLAB implementation may be written as

```
format longEng

Rs = 7e3
Rl = 2e3
f = 70.02817e3
w = f*2*pi

syms C1 L2 C3 L4 C5

E0 = ((Rs/Rl)+1)

E1 = w^1*(C1*Rs+L2/Rl+C3*Rs+L4/Rl+C5*Rs) ==
E0*(cos(0)/sin((1*pi)/10))

E2 = w^2*(C1*L2*Rs/Rl+C1*L4*Rs/Rl+C3*L2+C3*L4*Rs/Rl+C5*L2+C5*L4) ==
   E0*(cos(0)/sin((1*pi)/10))*
   (cos((1*pi)/10)/sin((2*pi)/10))

E3 = w^3*(C1*L2*C3*Rs+C1*L2*C5*Rs+C1*L4*C5*Rs+C3*L4*C5*Rs+(L2*C3*L4)/Rl) ==
   E0*(cos(0)/sin((1*pi)/10))*
   (cos((1*pi)/10)/sin((2*pi)/10))*
   (cos((2*pi)/10)/sin((3*pi)/10))

E4 = w^4*(C1*L2*C3*L4*Rs/Rl+L2*C3*L4*C5) ==
   E0*(cos(0)/sin((1*pi)/10))*
   (cos((1*pi)/10)/sin((2*pi)/10))*
   (cos((2*pi)/10)/sin((3*pi)/10))*
   (cos((3*pi)/10)/sin((4*pi)/10))

E5 = w^5*(C1*L2*C3*L4*C5*Rs) ==
   E0*(cos(0)/sin((1*pi)/10))*
   (cos((1*pi)/10)/sin((2*pi)/10))*
   (cos((2*pi)/10)/sin((3*pi)/10))*
   (cos((3*pi)/10)/sin((4*pi)/10))*
   (cos((4*pi)/10)/sin((5*pi)/10))

result = vpasolve(E1,E2,E3,E4,E5)

C1 = vpa(result.C1)
L2 = vpa(result.L2)
C3 = vpa(result.C3)
L4 = vpa(result.L4)
C5 = vpa(result.C5)

M_C1 = abs(C1)
M_L2 = abs(L2)
M_C3 = abs(C3)
M_L4 = abs(L4)
M_C5 = abs(C5)
```

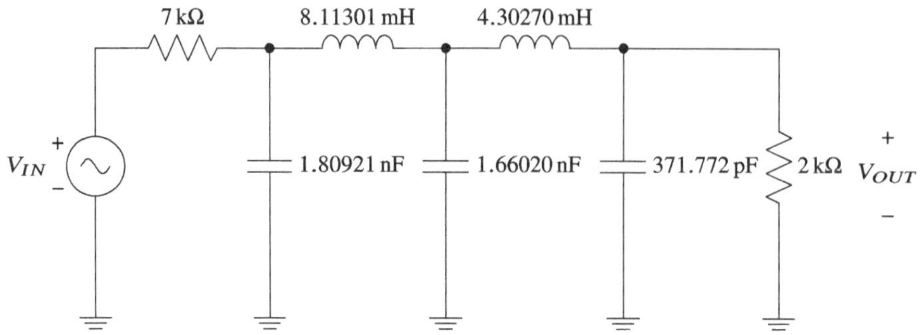

Fig. 10.5 Representative fifth-order shunt-first low-pass precursor for high-pass transformation

The complete candidate solution tuple output for the low-pass precursor solve used in this high-pass design example is provided in Appendix B, Sect. B.15.

For this low-pass precursor solve, the solver returns sixteen candidate tuples: four fully real tuples and six complex-conjugate solution pairs. Each complex-conjugate pair also produces a magnitude tuple. These magnitude tuples are less accurate than the fully real tuples and are not used as the representative passive realization. The four fully real tuples are retained as valid passive realizations.

To provide a concrete component set for the high-pass transformation, one of the fully real tuples is selected as a representative low-pass precursor. The fourth fully real tuple is used here for the transformation and verification steps (Fig. 10.5):

$$C_1 = 1.809210719835\,\text{nF}, \quad L_2 = 8.11301810866879\,\text{mH}, \quad C_3 = 1.6602010702419\,\text{nF},$$
$$L_4 = 4.30270708\,\text{mH}, \quad C_5 = 371.77242643469\,\text{pF}.$$

At this point, the synthesis step is complete; the remaining calculations apply the high-pass transformation to each reactive element.

All three low-pass shunt capacitors (C_1, C_3, and C_5) get transformed to high-pass shunt inductors (L_1, L_3, and L_5), and both low-pass series inductors (L_2 and L_4) get transformed to high-pass series capacitors (C_2 and C_4) using Eqs. (6.19) and (6.20).

The first high-pass shunt inductor, L_1, which is derived from C_1, becomes:

$$2.854997\,\text{mH} = \frac{1}{\left(440 \times 10^3 \frac{r}{s}\right)^2 * (1.80921\,\text{nF})}$$

The first high-pass series capacitor, C_2, which is derived from L_2, becomes:

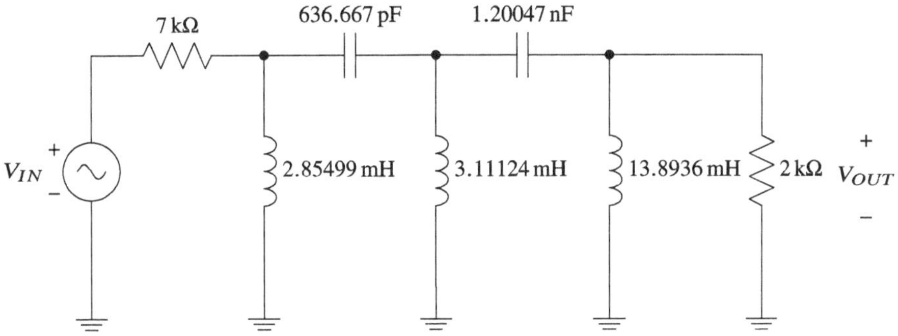

Fig. 10.6 Representative fifth-order shunt-first high-pass solution

$$636.667\,\text{pF} = \frac{1}{\left(440 \times 10^3 \frac{r}{s}\right)^2 * (8.113018\,\text{mH})}$$

The second high-pass shunt inductor, L_3, which is derived from C_3, becomes:

$$3.111243\,\text{mH} = \frac{1}{\left(440 \times 10^3 \frac{r}{s}\right)^2 * (1.660201\,\text{nF})}$$

The second high-pass series capacitor, C_4, which is derived from L_4, becomes:

$$1.200474\,\text{nF} = \frac{1}{\left(440 \times 10^3 \frac{r}{s}\right)^2 * (4.302707\,\text{mH})}$$

The third high-pass shunt inductor, L_5, which is derived from C_5, becomes:

$$13.89368\,\text{mH} = \frac{1}{\left(440 \times 10^3 \frac{r}{s}\right)^2 * (371.772426\,\text{pF})}$$

Producing a fifth-order shunt-first high-pass Butterworth filter (Fig. 10.6).

10.6 High-Pass Series-First

A fifth-order series-first high-pass filter is obtained by first synthesizing the corresponding fifth-order series-first low-pass ladder and then converting each reactive element using Eqs. (6.19) and (6.20). Therefore, the high-pass example is not a separate coefficient-matching derivation. It is a transformation example that uses the series-first low-pass system of Eq. (10.11) to obtain a low-pass precursor, which is then converted into the final high-pass ladder. Low-pass series inductors become high-pass series capacitors, and low-pass shunt capacitors become high-pass shunt inductors.

10.6.1 Design Example Fifth-Order High-Pass Series-First

The design requirements are:

- $R_S = 1\,\text{k}\Omega$
- $R_L = 4\,\text{k}\Omega$
- $f_c = 3.183098\,\text{kHz}$
- $\omega_c = 20 \times 10^3\,\text{rad/s}$

The passband gain established by the terminations is

$$20\log_{10}\left(\frac{4\,\text{k}\Omega}{4\,\text{k}\Omega + 1\,\text{k}\Omega}\right) = -1.9382\,\text{dB}.$$

Therefore the gain at the high-pass cut-on frequency is

$$20\log_{10}\left(\frac{1}{\sqrt{2}} * \frac{4\,\text{k}\Omega}{4\,\text{k}\Omega + 1\,\text{k}\Omega}\right) = -4.9485\,\text{dB},$$

or

$$-1.9382\,\text{dB} - 3.010299\,\text{dB} = -4.9485\,\text{dB}.$$

To obtain the low-pass precursor for the high-pass realization, the series-first low-pass system of Eq. 10.11 is solved using the high-pass design requirements. A MATLAB implementation may be written as

```
format longENG

Rs = 1e3
Rl = 4e3
f = 3.183098e3
w = f*2*pi


syms L1 C2 L3 C4 L5

E0 = ((Rs/Rl)+1)

E1 = w^1*(L1/Rl + C2*Rs + L3/Rl + C4*Rs + L5/Rl) ==
E0*(cos(0)/sin((1*pi)/10))

E2 = w^2*(L1*C2 + L1*C4 + L3*C2*Rs/Rl + L3*C4 + L5*C2*Rs/Rl + L5*C4*Rs/Rl) ==
E0*(cos(0)/sin((1*pi)/10))*
(cos((1*pi)/10)/sin((2*pi)/10))

E3 = w^3*(L1*C2*L3/Rl + L1*C2*L5/Rl + L1*C4*L5/Rl + C2*L3*C4*Rs + L3*C4*L5/Rl) ==
E0*(cos(0)/sin((1*pi)/10))*
(cos((1*pi)/10)/sin((2*pi)/10))*
(cos((2*pi)/10)/sin((3*pi)/10))

E4 = w^4*(L1*C2*L3*C4+C2*L3*C4*L5*Rs/Rl) ==
E0*(cos(0)/sin((1*pi)/10))*
(cos((1*pi)/10)/sin((2*pi)/10))*
```

```
(cos((2*pi)/10)/sin((3*pi)/10))*
(cos((3*pi)/10)/sin((4*pi)/10))

E5 = w^5*(L1*C2*L3*C4*L5)/R1 ==
E0*(cos(0)/sin((1*pi)/10))*
(cos((1*pi)/10)/sin((2*pi)/10))*
(cos((2*pi)/10)/sin((3*pi)/10))*
(cos((3*pi)/10)/sin((4*pi)/10))*
(cos((4*pi)/10)/sin((5*pi)/10))

result = vpasolve(E1,E2,E3,E4,E5)

L1 = vpa(result.L1)
C2 = vpa(result.C2)
L3 = vpa(result.L3)
C4 = vpa(result.C4)
L5 = vpa(result.L5)


M_L1 = abs(L1)
M_C2 = abs(C2)
M_L3 = abs(L3)
M_C4 = abs(C4)
M_L5 = abs(L5)
```

The complete candidate solution tuple output for the low-pass precursor solve used in this high-pass design example is provided in Appendix B, Sect. B.16.

For this low-pass precursor solve, the solver returns sixteen candidate tuples: four fully real tuples and six complex-conjugate solution pairs. Each complex-conjugate pair also produces a magnitude tuple. These magnitude tuples are less accurate than the fully real tuples and are not used as the representative passive realization. The four fully real tuples are retained as valid passive realizations.

To provide a concrete component set for the high-pass transformation, one of the fully real tuples is selected as a representative low-pass precursor. The first fully real tuple is used here for the transformation and verification steps (Fig. 10.7):

$$L_1 = 318.18216069\,\text{mH}, \quad C_2 = 22.176397394\,\text{nF}, \quad L_3 = 290.18076777\,\text{mH},$$
$$C_4 = 11.7477365313\,\text{nF}, \quad L_5 = 64.9577492495\,\text{mH}.$$

At this point, the synthesis step is complete; the remaining calculations apply the high-pass transformation to each reactive element.

All three low-pass series inductors (L_1, L_3, and L_5) get transformed to high-pass series capacitors (C_1, C_3, and C_5), and both low-pass shunt capacitors (C_2 and C_4) get transformed to high-pass shunt inductors (L_2 and L_4) using Eqs. (6.19) and (6.20).

The first high-pass series capacitor, C_1, which is derived from L_1, becomes:

$$7.857134\,\text{nF} = \frac{1}{(20 \times 10^3 \frac{r}{s})^2 * (318.18216069\,\text{mH})}$$

The first high-pass shunt inductor, L_2, which is derived from C_2, becomes:

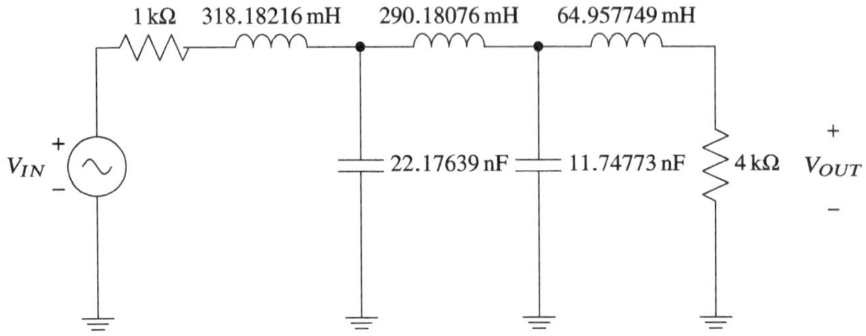

Fig. 10.7 Representative fifth-order series-first low-pass precursor for high-pass transformation

$$112.7324\,\text{mH} = \frac{1}{(20 \times 10^3 \frac{r}{s})^2 * (22.176397394\,\text{nF})}$$

The second high-pass series capacitor, C_3, which is derived from L_3, becomes:

$$8.615319\,\text{nF} = \frac{1}{(20 \times 10^3 \frac{r}{s})^2 * (290.18076777\,\text{mH})}$$

The second high-pass shunt inductor, L_4, which is derived from C_4, becomes:

$$212.8069\,\text{mH} = \frac{1}{(20 \times 10^3 \frac{r}{s})^2 * (11.7477365313\,\text{nF})}$$

The third high-pass series capacitor, C_5, which is derived from L_5, becomes:

$$38.48655\,\text{nF} = \frac{1}{(20 \times 10^3 \frac{r}{s})^2 * (64.9577492495\,\text{mH})}$$

Producing a fifth-order series-first high-pass Butterworth filter (Fig. 10.8).

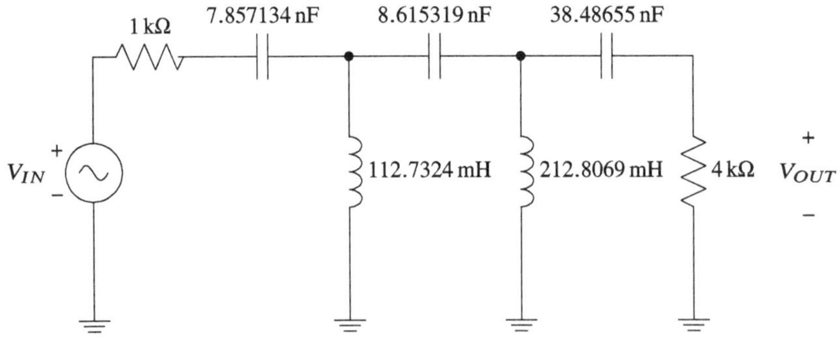

Fig. 10.8 Representative fifth-order series-first high-pass solution

10.7 Conclusion

The fifth-order case extends the exact synthesis method while preserving odd-order denominator symmetry. The fifth-degree term pairs with the constant term, the fourth-degree term pairs with the first-degree term, and the third-degree term pairs with the second-degree term.

The more important development occurs at the product-factor level. Fifth order introduces the first secondary correlated factor family. Because this secondary factor evaluates to unity, the normalized polynomial still simplifies to the same compact fifth-order coefficient structure represented by the equivalent entries in Table 10.1. Fifth order therefore acts as a transition case: it preserves the elegance of the lower-order relationships while revealing the structural feature that becomes unavoidable at sixth order.

The design examples show that the fifth-order synthesis equations still produce valid passive ladder realizations. The real-valued tuples remain usable circuit solutions, so the increased candidate-solution complexity should not be interpreted as a failure of the method. Instead, fifth order identifies the point where the product-factor transition and the solver-branch structure begin to change together.

The next chapter extends the same synthesis framework to sixth order. At sixth order, the secondary correlated factor family no longer collapses to unity; it appears explicitly as a reciprocal pair. This makes the sixth-order case the first order where the added product-factor structure becomes visible in the normalized denominator and where the computational burden begins to show clear diminishing returns.

References

1. Butterworth, S. (1930). On the theory of filter amplifiers. *Experimental Wireless and the Wireless Engineer, 7*, 536–541.
2. Horowitz, P., & Hill, W. (2015). *The art of electronics* (3rd ed.). Cambridge University Press. ISBN 978-0-521-80926-9 Hardback.
3. Zverev, A. I. (1967). *Handbook of filter synthesis*. Wiley.
4. Van Valkenburg, M. E. (1982). *Analog filter design*. Holt, Rinehart and Winston.
5. Schaumann, R., & Van Valkenburg, M.E. (2001). *Design of analog filters*. Oxford University Press.
6. Huelsman, L. P. (1993). *Active and passive analog filter design*. McGraw-Hill.

11 6th-Order Doubly-Terminated Passive Butterworth Filter Network Synthesis

Sixth order is where the hidden factor family becomes visible.

Chapter 10 extended the odd-order synthesis pattern to fifth order and showed where the product-factor structure first begins to change. In fifth order, the secondary correlated factor family appears through a unity factor. Because this factor evaluates to one, it does not visibly alter the normalized fifth-order coefficient structure. The sixth-order case makes that hidden structure visible.

As in the preceding chapters, the method itself does not change. Each degree-k frequency-dependent term of the realized ladder denominator is set equal, at cutoff, to the frequency-independent term multiplied by the corresponding normalized Butterworth coefficient. What changes at sixth order is the factor-level organization of those coefficients. The outer factor lines remain part of the primary correlated factor family, while the interior factor lines now form a visible secondary correlated factor family.

The Butterworth coefficient structure and ladder-network setting remain grounded in Butterworth's original maximally flat response [1] and the broader classical filter-synthesis literature [2]. The secondary factor-family interpretation and corresponding time-constant matching are developed here as part of the present construction.

K. M. P. Ryan, *Polynomial Symmetry in Butterworth Filter Synthesis*,
Synthesis Lectures on Engineering, Science, and Technology,
https://doi.org/10.1007/978-3-032-33468-8_11

11.1 Even Symmetry

As with the second- and fourth-order cases, sixth-order doubly-terminated passive Butterworth filters are even-order and therefore topologically constrained. A shunt-first circuit topology is used when $R_S > R_L$, and a series-first circuit topology is used when $R_S < R_L$. When $R_S = R_L$, either topology may be used.

The sixth-order synthesis problem produces a larger candidate solution set than the lower-order cases. In the present work, fully real solutions are retained as valid passive realizations. Solutions containing imaginary components may still be algebraically informative, but they are not treated as exact passive filter designs.

At the denominator level, sixth order follows the expected even-order pattern: the sixth-degree term is paired with the constant term, the fifth- and first-degree terms are paired, the fourth- and second-degree terms are paired, and the third-degree term stands alone. At the factor level, however, the third-degree coefficient is where the secondary correlated factor family becomes visible.

Table 11.1 gives the sixth-order Butterworth product factors in chapter context. The complete order-by-order comparison remains collected in Appendix A.

Lines 1, 2, 5, and 6 of Table 11.1 remain mutually expressible through the primary scaling and inversion relationships. Lines 3 and 4 form the secondary correlated factor family. They are related to each other through inversion,

$$\sqrt{\frac{3}{2}} \quad \text{and} \quad \frac{1}{\sqrt{\frac{3}{2}}},$$

but they are not generated by the same primary relationship that connects the outer factor lines. This is the first order in which the secondary correlated factor family is visible as a non-unity reciprocal pair.

The factors in Table 11.1 are multiplied cumulatively to form the normalized sixth-order denominator polynomial. Written as a denominator function, this gives

$$\begin{aligned} D_6(s) = {} & s^6 + (\text{Line } 1)s^5 + (\text{Line } 1)(\text{Line } 2)s^4 \\ & + (\text{Line } 1)(\text{Line } 2)(\text{Line } 3)s^3 \\ & + (\text{Line } 1)(\text{Line } 2)(\text{Line } 3)(\text{Line } 4)s^2 \\ & + (\text{Line } 1)(\text{Line } 2)(\text{Line } 3)(\text{Line } 4)(\text{Line } 5)s \\ & + (\text{Line } 1)(\text{Line } 2)(\text{Line } 3)(\text{Line } 4)(\text{Line } 5)(\text{Line } 6), \end{aligned}$$

where any equivalent entry from the corresponding table line may be substituted. Using the numerical factors,

Table 11.1 Sixth-order Butterworth polynomial factors with respect to Eq. (2.1) (variations 1–6), and derivations (variations 7–10)

Line	Factor	Variation 1	Variation 2	Variation 3	Variation 4	Variation 5	Variation 6	Variation 7	Variation 8	Variation 9	Variation 10
1	3.863703	$\left(\frac{\cos(0)}{\sin\left(\frac{1}{12}\pi\right)}\right)$	$2\left(\frac{\cos\left(\frac{1}{12}\pi\right)}{\sin\left(\frac{2}{12}\pi\right)}\right)$			$\frac{2}{\left(\frac{\cos\left(\frac{4}{12}\pi\right)}{\sin\left(\frac{5}{12}\pi\right)}\right)}$	$\frac{1}{\left(\frac{\cos\left(\frac{5}{12}\pi\right)}{\sin\left(\frac{6}{12}\pi\right)}\right)}$	$\left(2(\sqrt{2+\sqrt{3}})\right)$	$\frac{2}{\sqrt{2-\sqrt{3}}}$		$\csc\left(\frac{\pi}{12}\right)$
2	1.931851	$\frac{\left(\frac{\cos(0)}{\sin\left(\frac{1}{12}\pi\right)}\right)}{2}$	$\left(\frac{\cos\left(\frac{1}{12}\pi\right)}{\sin\left(\frac{2}{12}\pi\right)}\right)$			$\frac{1}{\left(\frac{\cos\left(\frac{4}{12}\pi\right)}{\sin\left(\frac{5}{12}\pi\right)}\right)}$	$\frac{1}{2\left(\frac{\cos\left(\frac{5}{12}\pi\right)}{\sin\left(\frac{6}{12}\pi\right)}\right)}$	$\left(\left(\sqrt{2+\sqrt{3}}\right)\right)$	$\frac{1}{\sqrt{2-\sqrt{3}}}$		$\frac{\csc\left(\frac{\pi}{12}\right)}{2}$
3	1.224744			$\left(\frac{\cos\left(\frac{2}{12}\pi\right)}{\sin\left(\frac{3}{12}\pi\right)}\right)$	$\frac{1}{\left(\frac{\cos\left(\frac{3}{12}\pi\right)}{\sin\left(\frac{4}{12}\pi\right)}\right)}$					$\sqrt{\frac{3}{2}}$	
4	0.8164965			$\frac{1}{\left(\frac{\cos\left(\frac{2}{12}\pi\right)}{\sin\left(\frac{3}{12}\pi\right)}\right)}$	$\left(\frac{\cos\left(\frac{3}{12}\pi\right)}{\sin\left(\frac{4}{12}\pi\right)}\right)$					$\frac{1}{\sqrt{\frac{3}{2}}}$	
5	0.5176380	$\frac{2}{\left(\frac{\cos(0)}{\sin\left(\frac{1}{12}\pi\right)}\right)}$	$\frac{1}{\left(\frac{\cos\left(\frac{1}{12}\pi\right)}{\sin\left(\frac{2}{12}\pi\right)}\right)}$			$\left(\frac{\cos\left(\frac{4}{12}\pi\right)}{\sin\left(\frac{5}{12}\pi\right)}\right)$	$2\left(\frac{\cos\left(\frac{5}{12}\pi\right)}{\sin\left(\frac{6}{12}\pi\right)}\right)$	$\left(\frac{1}{\left(\sqrt{2+\sqrt{3}}\right)}\right)$	$\sqrt{2-\sqrt{3}}$		$\frac{2}{\csc\left(\frac{\pi}{12}\right)}$
6	0.2588190	$\frac{1}{\left(\frac{\cos(0)}{\sin\left(\frac{1}{12}\pi\right)}\right)}$	$\frac{1}{2\left(\frac{\cos\left(\frac{1}{12}\pi\right)}{\sin\left(\frac{2}{12}\pi\right)}\right)}$			$\frac{\left(\frac{\cos\left(\frac{4}{12}\pi\right)}{\sin\left(\frac{5}{12}\pi\right)}\right)}{2}$	$\left(\frac{\cos\left(\frac{5}{12}\pi\right)}{\sin\left(\frac{6}{12}\pi\right)}\right)$	$\left(\frac{1}{2\left(\sqrt{2+\sqrt{3}}\right)}\right)$	$\frac{\sqrt{2-\sqrt{3}}}{2}$		$\frac{1}{\csc\left(\frac{\pi}{12}\right)}$

$$\begin{aligned}
D_6(s) = s^6 &+ (3.863703)s^5 + (3.863703)(1.931851)s^4 \\
&+ (3.863703)(1.931851)(1.224744)s^3 \\
&+ (3.863703)(1.931851)(1.224744)(0.8164965)s^2 \\
&+ (3.863703)(1.931851)(1.224744)(0.8164965)(0.5176380)s \\
&+ (3.863703)(1.931851)(1.224744)(0.8164965)(0.5176380)(0.2588190).
\end{aligned}$$

This explicit product form is important in the sixth-order case because the middle reciprocal pair is not produced by the outer primary family alone. The denominator coefficient values can still be written compactly, but the factor-level structure is no longer exhausted by the primary correlated factor family.

11.2 Diminishing Returns in Sixth-Order Solution Search

The preceding chapters included representative design examples because the lower-order systems were still small enough for the examples to be useful without obscuring the main synthesis pattern. At sixth order, the situation changes. The algebraic system contains six unknown reactive element values, and the candidate solution set expands enough that exhaustive example generation becomes computationally expensive relative to the additional insight it provides.

This is not a change in the synthesis method. The same degree-matching rule still applies, and real-valued solution tuples remain valid passive realizations. The issue is practical: once the sixth-order factor structure is visible, additional numerical examples tend to repeat the same interpretation while requiring substantially more computation.

Table 11.1 gives a plausible reason for this behavior. Lines 1, 2, 5, and 6 remain connected through the primary correlated factor family, while Lines 3 and 4 form the secondary correlated factor family. The appearance of this reciprocal middle pair does not by itself prove the source of long symbolic or MATLAB compilation times, but it does show that the sixth-order product-factor structure is no longer governed by a single correlated family. For that reason, the remainder of this chapter emphasizes the factor-family structure, the degree-matching equations, and the resulting time-constant identities rather than adding a long sequence of sixth-order numerical design examples.

11.3 Low-Pass Shunt-First

The transfer function for the sixth-order shunt-first doubly-terminated low-pass filter in Fig. 11.1 is

$$\begin{aligned}\frac{V_{OUT}}{V_{IN}} = \frac{1}{\begin{aligned}&\omega^6 C_1L_2C_3L_4C_5L_6\frac{R_S}{R_L}\\&+\omega^5\left(C_1L_2C_3L_4C_5R_S + L_2C_3L_4C_5L_6\frac{1}{R_L}\right)\\&+\omega^4\left(C_1L_2C_3L_4\frac{R_S}{R_L} + C_1L_2C_3L_6\frac{R_S}{R_L} + C_1L_2C_5L_6\frac{R_S}{R_L}\right.\\&\qquad\left. + C_1L_4C_5L_6\frac{R_S}{R_L} + C_3L_4C_5L_6\frac{R_S}{R_L} + L_2C_3L_4C_5\right)\\&+\omega^3\left(C_1L_2C_3R_S + C_1L_2C_5R_S + C_1L_4C_5R_S + C_3L_4C_5R_S\right.\\&\qquad\left. + L_2C_3L_4\frac{1}{R_L} + L_2C_3L_6\frac{1}{R_L} + L_2C_5L_6\frac{1}{R_L} + L_4C_5L_6\frac{1}{R_L}\right)\\&+\omega^2\left(C_1L_2\frac{R_S}{R_L} + C_1L_4\frac{R_S}{R_L} + C_1L_6\frac{R_S}{R_L} + C_3L_2\right.\\&\qquad\left. + C_3L_4\frac{R_S}{R_L} + C_3L_6\frac{R_S}{R_L} + C_5L_2 + C_5L_4 + C_5L_6\frac{R_S}{R_L}\right)\\&+\omega\left(C_1R_S + \frac{L_2}{R_L} + C_3R_S + \frac{L_4}{R_L} + C_5R_S + \frac{L_6}{R_L}\right) + \left(\frac{R_S}{R_L} + 1\right)\end{aligned}}\end{aligned} \tag{11.1}$$

The corresponding Butterworth transfer function using the product formula of Eq. (2.1) is

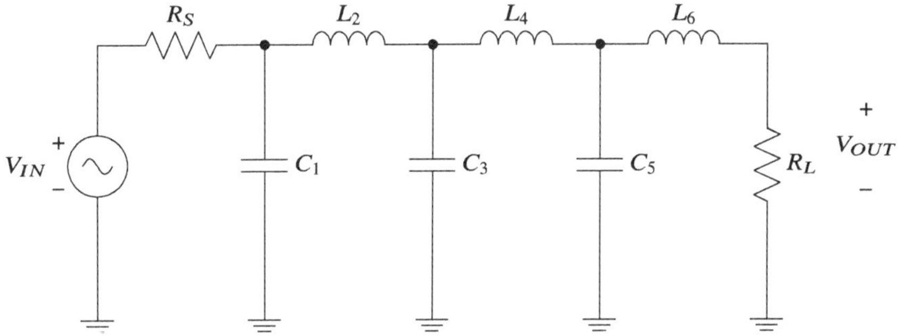

Fig. 11.1 Sixth-order shunt-first doubly-terminated low-pass Butterworth filter

$$\frac{V_{OUT}}{V_{IN}} = \frac{1}{\begin{aligned} & s^6\left(\left(\frac{\cos(0)}{\sin\left(\frac{1}{12}\pi\right)}\right)\left(\frac{\cos(\frac{1}{12}\pi)}{\sin\left(\frac{2}{12}\pi\right)}\right)\left(\frac{\cos(\frac{2}{12}\pi)}{\sin\left(\frac{3}{12}\pi\right)}\right)\right. \\ & \left.\left(\frac{\cos(\frac{3}{12}\pi)}{\sin\left(\frac{4}{12}\pi\right)}\right)\left(\frac{\cos(\frac{4}{12}\pi)}{\sin\left(\frac{5}{12}\pi\right)}\right)\left(\frac{\cos(\frac{5}{12}\pi)}{\sin\left(\frac{6}{12}\pi\right)}\right)\right) \\ & + s^5\left(\left(\frac{\cos(0)}{\sin\left(\frac{1}{12}\pi\right)}\right)\left(\frac{\cos(\frac{1}{12}\pi)}{\sin\left(\frac{2}{12}\pi\right)}\right)\left(\frac{\cos(\frac{2}{12}\pi)}{\sin\left(\frac{3}{12}\pi\right)}\right)\right. \\ & \left.\left(\frac{\cos(\frac{3}{12}\pi)}{\sin\left(\frac{4}{12}\pi\right)}\right)\left(\frac{\cos(\frac{4}{12}\pi)}{\sin\left(\frac{5}{12}\pi\right)}\right)\right) \\ & + s^4\left(\left(\frac{\cos(0)}{\sin\left(\frac{1}{12}\pi\right)}\right)\left(\frac{\cos(\frac{1}{12}\pi)}{\sin\left(\frac{2}{12}\pi\right)}\right)\left(\frac{\cos(\frac{2}{12}\pi)}{\sin\left(\frac{3}{12}\pi\right)}\right)\left(\frac{\cos(\frac{3}{12}\pi)}{\sin\left(\frac{4}{12}\pi\right)}\right)\right) \\ & + s^3\left(\left(\frac{\cos(0)}{\sin\left(\frac{1}{12}\pi\right)}\right)\left(\frac{\cos(\frac{1}{12}\pi)}{\sin\left(\frac{2}{12}\pi\right)}\right)\left(\frac{\cos(\frac{2}{12}\pi)}{\sin\left(\frac{3}{12}\pi\right)}\right)\right) \\ & + s^2\left(\left(\frac{\cos(0)}{\sin\left(\frac{1}{12}\pi\right)}\right)\left(\frac{\cos(\frac{1}{12}\pi)}{\sin\left(\frac{2}{12}\pi\right)}\right)\right) \\ & + s\left(\frac{\cos(0)}{\sin\left(\frac{1}{12}\pi\right)}\right) + 1 \end{aligned}} \tag{11.2}$$

The sixth-degree term of the realized denominator,

$$\omega^6 C_1 L_2 C_3 L_4 C_5 L_6 \frac{R_S}{R_L},$$

is associated with the sixth-degree Butterworth term

$$\begin{aligned} s^6 & \left(\left(\frac{\cos(0)}{\sin\left(\frac{1}{12}\pi\right)}\right)\left(\frac{\cos\left(\frac{1}{12}\pi\right)}{\sin\left(\frac{2}{12}\pi\right)}\right)\left(\frac{\cos\left(\frac{2}{12}\pi\right)}{\sin\left(\frac{3}{12}\pi\right)}\right)\right. \\ & \left.\left(\frac{\cos\left(\frac{3}{12}\pi\right)}{\sin\left(\frac{4}{12}\pi\right)}\right)\left(\frac{\cos\left(\frac{4}{12}\pi\right)}{\sin\left(\frac{5}{12}\pi\right)}\right)\left(\frac{\cos\left(\frac{5}{12}\pi\right)}{\sin\left(\frac{6}{12}\pi\right)}\right)\right). \end{aligned}$$

The fifth-degree term of the realized denominator,

$$\omega^5\left(C_1 L_2 C_3 L_4 C_5 R_S + L_2 C_3 L_4 C_5 L_6 \frac{1}{R_L}\right),$$

is associated with the fifth-degree Butterworth term

$$s^5\left(\left(\frac{\cos(0)}{\sin\left(\frac{1}{12}\pi\right)}\right)\left(\frac{\cos\left(\frac{1}{12}\pi\right)}{\sin\left(\frac{2}{12}\pi\right)}\right)\left(\frac{\cos\left(\frac{2}{12}\pi\right)}{\sin\left(\frac{3}{12}\pi\right)}\right)\left(\frac{\cos\left(\frac{3}{12}\pi\right)}{\sin\left(\frac{4}{12}\pi\right)}\right)\left(\frac{\cos\left(\frac{4}{12}\pi\right)}{\sin\left(\frac{5}{12}\pi\right)}\right)\right).$$

The fourth-degree term of the realized denominator,

$$\omega^4\left(C_1L_2C_3L_4\frac{R_S}{R_L}+C_1L_2C_3L_6\frac{R_S}{R_L}+C_1L_2C_5L_6\frac{R_S}{R_L}\right.$$
$$\left.+C_1L_4C_5L_6\frac{R_S}{R_L}+C_3L_4C_5L_6\frac{R_S}{R_L}+L_2C_3L_4C_5\right),$$

is associated with the fourth-degree Butterworth term

$$s^4\left(\left(\frac{\cos(0)}{\sin\left(\frac{1}{12}\pi\right)}\right)\left(\frac{\cos\left(\frac{1}{12}\pi\right)}{\sin\left(\frac{2}{12}\pi\right)}\right)\left(\frac{\cos\left(\frac{2}{12}\pi\right)}{\sin\left(\frac{3}{12}\pi\right)}\right)\left(\frac{\cos\left(\frac{3}{12}\pi\right)}{\sin\left(\frac{4}{12}\pi\right)}\right)\right).$$

The third-degree term of the realized denominator,

$$\omega^3\left(C_1L_2C_3R_S+C_1L_2C_5R_S+C_1L_4C_5R_S+C_3L_4C_5R_S\right.$$
$$\left.+L_2C_3L_4\frac{1}{R_L}+L_2C_3L_6\frac{1}{R_L}+L_2C_5L_6\frac{1}{R_L}+L_4C_5L_6\frac{1}{R_L}\right),$$

is associated with the third-degree Butterworth term

$$s^3\left(\left(\frac{\cos(0)}{\sin\left(\frac{1}{12}\pi\right)}\right)\left(\frac{\cos\left(\frac{1}{12}\pi\right)}{\sin\left(\frac{2}{12}\pi\right)}\right)\left(\frac{\cos\left(\frac{2}{12}\pi\right)}{\sin\left(\frac{3}{12}\pi\right)}\right)\right).$$

The second-degree term of the realized denominator,

$$\omega^2\left(C_1L_2\frac{R_S}{R_L}+C_1L_4\frac{R_S}{R_L}+C_1L_6\frac{R_S}{R_L}+C_3L_2\right.$$
$$\left.+C_3L_4\frac{R_S}{R_L}+C_3L_6\frac{R_S}{R_L}+C_5L_2+C_5L_4+C_5L_6\frac{R_S}{R_L}\right),$$

is associated with the second-degree Butterworth term

$$s^2\left(\left(\frac{\cos(0)}{\sin\left(\frac{1}{12}\pi\right)}\right)\left(\frac{\cos\left(\frac{1}{12}\pi\right)}{\sin\left(\frac{2}{12}\pi\right)}\right)\right).$$

The first-degree term of the realized denominator,

$$\omega\left(C_1R_S+\frac{L_2}{R_L}+C_3R_S+\frac{L_4}{R_L}+C_5R_S+\frac{L_6}{R_L}\right),$$

is associated with the first-degree Butterworth term

$$s\left(\left(\frac{\cos(0)}{\sin\left(\frac{1}{12}\pi\right)}\right)\right).$$

Applying the transformation rule yields

$$\begin{aligned}
&AC_{aspect} = \left(DC_{aspect}\right)\left(Butterworth_{coef}\right)\\
&\omega^6 C_1 L_2 C_3 L_4 C_5 L_6 \frac{R_S}{R_L} = \left(\frac{R_S}{R_L}+1\right)\left(\left(\frac{\cos(0)}{\sin\left(\frac{1}{12}\pi\right)}\right)\left(\frac{\cos\left(\frac{1}{12}\pi\right)}{\sin\left(\frac{2}{12}\pi\right)}\right)\right.\\
&\left.\left(\frac{\cos\left(\frac{2}{12}\pi\right)}{\sin\left(\frac{3}{12}\pi\right)}\right)\left(\frac{\cos\left(\frac{3}{12}\pi\right)}{\sin\left(\frac{4}{12}\pi\right)}\right)\left(\frac{\cos\left(\frac{4}{12}\pi\right)}{\sin\left(\frac{5}{12}\pi\right)}\right)\left(\frac{\cos\left(\frac{5}{12}\pi\right)}{\sin\left(\frac{6}{12}\pi\right)}\right)\right)\\
&\omega^5\left(C_1 L_2 C_3 L_4 C_5 R_S + L_2 C_3 L_4 C_5 L_6 \frac{1}{R_L}\right)\\
&= \left(\frac{R_S}{R_L}+1\right)\left(\left(\frac{\cos(0)}{\sin\left(\frac{1}{12}\pi\right)}\right)\left(\frac{\cos\left(\frac{1}{12}\pi\right)}{\sin\left(\frac{2}{12}\pi\right)}\right)\right.\\
&\left.\left(\frac{\cos\left(\frac{2}{12}\pi\right)}{\sin\left(\frac{3}{12}\pi\right)}\right)\left(\frac{\cos\left(\frac{3}{12}\pi\right)}{\sin\left(\frac{4}{12}\pi\right)}\right)\left(\frac{\cos\left(\frac{4}{12}\pi\right)}{\sin\left(\frac{5}{12}\pi\right)}\right)\right)\\
&\omega^4\left(C_1 L_2 C_3 L_4 \frac{R_S}{R_L} + C_1 L_2 C_3 L_6 \frac{R_S}{R_L} + C_1 L_2 C_5 L_6 \frac{R_S}{R_L}\right.\\
&\left. + C_1 L_4 C_5 L_6 \frac{R_S}{R_L} + C_3 L_4 C_5 L_6 \frac{R_S}{R_L} + L_2 C_3 L_4 C_5\right)\\
&= \left(\frac{R_S}{R_L}+1\right)\left(\left(\frac{\cos(0)}{\sin\left(\frac{1}{12}\pi\right)}\right)\left(\frac{\cos\left(\frac{1}{12}\pi\right)}{\sin\left(\frac{2}{12}\pi\right)}\right)\right.\\
&\left.\left(\frac{\cos\left(\frac{2}{12}\pi\right)}{\sin\left(\frac{3}{12}\pi\right)}\right)\left(\frac{\cos\left(\frac{3}{12}\pi\right)}{\sin\left(\frac{4}{12}\pi\right)}\right)\right)\\
&\omega^3\left(C_1 L_2 C_3 R_S + C_1 L_2 C_5 R_S + C_1 L_4 C_5 R_S + C_3 L_4 C_5 R_S\right.\\
&\left. + L_2 C_3 L_4 \frac{1}{R_L} + L_2 C_3 L_6 \frac{1}{R_L} + L_2 C_5 L_6 \frac{1}{R_L} + L_4 C_5 L_6 \frac{1}{R_L}\right)\\
&= \left(\frac{R_S}{R_L}+1\right)\left(\left(\frac{\cos(0)}{\sin\left(\frac{1}{12}\pi\right)}\right)\left(\frac{\cos\left(\frac{1}{12}\pi\right)}{\sin\left(\frac{2}{12}\pi\right)}\right)\right.\\
&\left.\left(\frac{\cos\left(\frac{2}{12}\pi\right)}{\sin\left(\frac{3}{12}\pi\right)}\right)\right)\\
&\omega^2\left(C_1 L_2 \frac{R_S}{R_L} + C_1 L_4 \frac{R_S}{R_L} + C_1 L_6 \frac{R_S}{R_L} + C_3 L_2\right.\\
&\left. + C_3 L_4 \frac{R_S}{R_L} + C_3 L_6 \frac{R_S}{R_L} + C_5 L_2 + C_5 L_4 + C_5 L_6 \frac{R_S}{R_L}\right)\\
&= \left(\frac{R_S}{R_L}+1\right)\left(\left(\frac{\cos(0)}{\sin\left(\frac{1}{12}\pi\right)}\right)\left(\frac{\cos\left(\frac{1}{12}\pi\right)}{\sin\left(\frac{2}{12}\pi\right)}\right)\right)\\
&\omega\left(C_1 R_S + \frac{L_2}{R_L} + C_3 R_S + \frac{L_4}{R_L} + C_5 R_S + \frac{L_6}{R_L}\right) = \left(\frac{R_S}{R_L}+1\right)\left(\left(\frac{\cos(0)}{\sin\left(\frac{1}{12}\pi\right)}\right)\right)
\end{aligned} \tag{11.3}$$

The system of Eq. (11.3) simplifies to

$$\omega^6 C_1 L_2 C_3 L_4 C_5 L_6 \frac{R_S}{R_L} = \left(\frac{R_S}{R_L} + 1\right), \tag{11.4}$$

$$\begin{aligned}\omega^5 \left(C_1 L_2 C_3 L_4 C_5 R_S + L_2 C_3 L_4 C_5 L_6 \frac{1}{R_L}\right) &= \left(2\sqrt{2+\sqrt{3}}\right)\left(\frac{R_S}{R_L}+1\right)\\ &= \left(\frac{2}{\sqrt{2-\sqrt{3}}}\right)\left(\frac{R_S}{R_L}+1\right)\\ &= \left(\csc\left(\frac{\pi}{12}\right)\right)\left(\frac{R_S}{R_L}+1\right),\end{aligned} \tag{11.5}$$

$$\begin{aligned}&\omega^4 \left(C_1 L_2 C_3 L_4 \frac{R_S}{R_L} + C_1 L_2 C_3 L_6 \frac{R_S}{R_L} + C_1 L_2 C_5 L_6 \frac{R_S}{R_L} + C_1 L_4 C_5 L_6 \frac{R_S}{R_L}\right.\\ &\quad \left. + C_3 L_4 C_5 L_6 \frac{R_S}{R_L} + L_2 C_3 L_4 C_5\right)\\ &= \left(2\left(2+\sqrt{3}\right)\right)\left(\frac{R_S}{R_L}+1\right)\\ &= \left(\frac{2}{2-\sqrt{3}}\right)\left(\frac{R_S}{R_L}+1\right)\\ &= \left(\frac{\csc(\frac{\pi}{12})^2}{2}\right)\left(\frac{R_S}{R_L}+1\right),\end{aligned} \tag{11.6}$$

$$\begin{aligned}&\omega^3 \left(C_1 L_2 C_3 R_S + C_1 L_2 C_5 R_S + C_1 L_4 C_5 R_S + C_3 L_4 C_5 R_S + L_2 C_3 L_4 \frac{1}{R_L} + L_2 C_3 L_6 \frac{1}{R_L}\right.\\ &\quad \left. + L_2 C_5 L_6 \frac{1}{R_L} + L_4 C_5 L_6 \frac{1}{R_L}\right)\\ &= \left(2\left(2+\sqrt{3}\right)\sqrt{\frac{3}{2}}\right)\left(\frac{R_S}{R_L}+1\right)\\ &= \left(\frac{\csc(\frac{\pi}{12})^2}{2}\sqrt{\frac{3}{2}}\right)\left(\frac{R_S}{R_L}+1\right)\\ &= \left(\sqrt{6}\left(2+\sqrt{3}\right)\right)\left(\frac{R_S}{R_L}+1\right),\end{aligned} \tag{11.7}$$

$$
\begin{aligned}
&\omega^2\left(C_1L_2\frac{R_S}{R_L}+C_1L_4\frac{R_S}{R_L}+C_1L_6\frac{R_S}{R_L}+C_3L_2+C_3L_4\frac{R_S}{R_L}\right.\\
&\left.+C_3L_6\frac{R_S}{R_L}+C_5L_2+C_5L_4+C_5L_6\frac{R_S}{R_L}\right)\\
&=\left(2\left(2+\sqrt{3}\right)\right)\left(\frac{R_S}{R_L}+1\right)\\
&=\left(\frac{2}{2-\sqrt{3}}\right)\left(\frac{R_S}{R_L}+1\right)\\
&=\left(\frac{\csc(\frac{\pi}{12})^2}{2}\right)\left(\frac{R_S}{R_L}+1\right),
\end{aligned}
\tag{11.8}
$$

and

$$
\begin{aligned}
\omega\left(C_1R_S+\frac{L_2}{R_L}+C_3R_S+\frac{L_4}{R_L}+C_5R_S+\frac{L_6}{R_L}\right)&=\left(2\sqrt{2+\sqrt{3}}\right)\left(\frac{R_S}{R_L}+1\right)\\
&=\left(\frac{2}{\sqrt{2-\sqrt{3}}}\right)\left(\frac{R_S}{R_L}+1\right)\\
&=\left(\csc\left(\frac{\pi}{12}\right)\right)\left(\frac{R_S}{R_L}+1\right).
\end{aligned}
\tag{11.9}
$$

The sixth-degree term and the constant term are equal. The fifth- and first-degree terms are equal, and the fourth- and second-degree terms are equal. The third-degree term stands alone. This is the sixth-order version of the even-order pattern seen in Chap. 9, now with the secondary correlated factor family visible in the middle coefficient.

Since there are six equations and six unknown reactive element values, the sixth-order system can produce multiple candidate solution tuples. In practice, the real-valued tuples are retained as exact passive realizations.

11.4 Low-Pass Series-First

For $R_S < R_L$, the complementary sixth-order low-pass topology is series-first, as shown in Fig. 11.2. This is the appropriate even-order realization when the load resistance exceeds the source resistance.

The transfer function for the sixth-order series-first doubly-terminated low-pass filter in Fig. 11.2 is

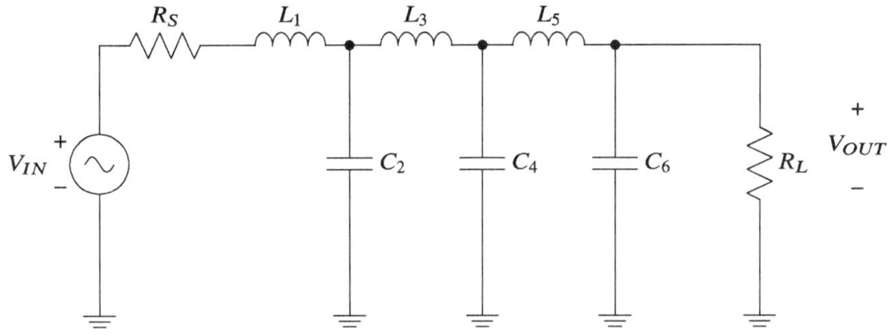

Fig. 11.2 Sixth-order series-first doubly-terminated low-pass Butterworth filter

$$\frac{V_{OUT}}{V_{IN}} = \frac{1}{\begin{aligned}&\omega^6 L_1C_2L_3C_4L_5C_6\\ &+\omega^5\left(C_2C_4C_6L_3L_5R_S + C_2C_4L_1L_3L_5\frac{1}{R_L}\right)\\ &+\omega^4\left(C_2C_4L_1L_3 + C_2C_4L_3L_5\frac{R_S}{R_L} + C_2C_6L_1L_3\right.\\ &\qquad\left. + C_2C_6L_1L_5 + C_4C_6L_1L_5 + C_4C_6L_3L_5\right)\\ &+\omega^3\left(C_2C_4L_3R_S + C_2C_6L_3R_S + C_2C_6L_5R_S\right.\\ &\qquad + C_2L_1L_3\frac{1}{R_L} + C_2L_1L_5\frac{1}{R_L} + C_4C_6L_5R_S\\ &\qquad\left. + C_4L_1L_5\frac{1}{R_L} + C_4L_3L_5\frac{1}{R_L}\right)\\ &+\omega^2\left(C_2L_1 + C_2L_3\frac{R_S}{R_L} + C_2L_5\frac{R_S}{R_L} + C_4L_1\right.\\ &\qquad\left. + C_4L_3 + C_4L_5\frac{R_S}{R_L} + C_6L_1 + C_6L_3 + C_6L_5\right)\\ &+\omega\left(\frac{L_1}{R_L} + C_2R_S + \frac{L_3}{R_L} + C_4R_S + \frac{L_5}{R_L} + C_6R_S\right) + \left(\frac{R_S}{R_L} + 1\right)\end{aligned}} \tag{11.10}$$

The corresponding Butterworth transfer function using the product formula of Eq. (2.1) is

$$\frac{V_{OUT}}{V_{IN}} = \frac{1}{\begin{aligned} & s^6\left(\left(\frac{\cos(0)}{\sin\left(\frac{1}{12}\pi\right)}\right)\left(\frac{\cos(\frac{1}{12}\pi)}{\sin\left(\frac{2}{12}\pi\right)}\right)\left(\frac{\cos(\frac{2}{12}\pi)}{\sin\left(\frac{3}{12}\pi\right)}\right)\right. \\ & \left.\left(\frac{\cos(\frac{3}{12}\pi)}{\sin\left(\frac{4}{12}\pi\right)}\right)\left(\frac{\cos(\frac{4}{12}\pi)}{\sin\left(\frac{5}{12}\pi\right)}\right)\left(\frac{\cos(\frac{5}{12}\pi)}{\sin\left(\frac{6}{12}\pi\right)}\right)\right) \\ & + s^5\left(\left(\frac{\cos(0)}{\sin\left(\frac{1}{12}\pi\right)}\right)\left(\frac{\cos(\frac{1}{12}\pi)}{\sin\left(\frac{2}{12}\pi\right)}\right)\left(\frac{\cos(\frac{2}{12}\pi)}{\sin\left(\frac{3}{12}\pi\right)}\right)\right. \\ & \left.\left(\frac{\cos(\frac{3}{12}\pi)}{\sin\left(\frac{4}{12}\pi\right)}\right)\left(\frac{\cos(\frac{4}{12}\pi)}{\sin\left(\frac{5}{12}\pi\right)}\right)\right) \\ & + s^4\left(\left(\frac{\cos(0)}{\sin\left(\frac{1}{12}\pi\right)}\right)\left(\frac{\cos(\frac{1}{12}\pi)}{\sin\left(\frac{2}{12}\pi\right)}\right)\left(\frac{\cos(\frac{2}{12}\pi)}{\sin\left(\frac{3}{12}\pi\right)}\right)\left(\frac{\cos(\frac{3}{12}\pi)}{\sin\left(\frac{4}{12}\pi\right)}\right)\right) \\ & + s^3\left(\left(\frac{\cos(0)}{\sin\left(\frac{1}{12}\pi\right)}\right)\left(\frac{\cos(\frac{1}{12}\pi)}{\sin\left(\frac{2}{12}\pi\right)}\right)\left(\frac{\cos(\frac{2}{12}\pi)}{\sin\left(\frac{3}{12}\pi\right)}\right)\right) \\ & + s^2\left(\left(\frac{\cos(0)}{\sin\left(\frac{1}{12}\pi\right)}\right)\left(\frac{\cos(\frac{1}{12}\pi)}{\sin\left(\frac{2}{12}\pi\right)}\right)\right) \\ & + s\left(\frac{\cos(0)}{\sin\left(\frac{1}{12}\pi\right)}\right) + 1 \end{aligned}} \tag{11.11}$$

The sixth-degree term of the realized denominator,

$$\omega^6 L_1 C_2 L_3 C_4 L_5 C_6,$$

is associated with the sixth-degree Butterworth term

$$s^6\left(\left(\frac{\cos(0)}{\sin\left(\frac{1}{12}\pi\right)}\right)\left(\frac{\cos\left(\frac{1}{12}\pi\right)}{\sin\left(\frac{2}{12}\pi\right)}\right)\left(\frac{\cos\left(\frac{2}{12}\pi\right)}{\sin\left(\frac{3}{12}\pi\right)}\right)\left(\frac{\cos\left(\frac{3}{12}\pi\right)}{\sin\left(\frac{4}{12}\pi\right)}\right)\left(\frac{\cos\left(\frac{4}{12}\pi\right)}{\sin\left(\frac{5}{12}\pi\right)}\right)\left(\frac{\cos\left(\frac{5}{12}\pi\right)}{\sin\left(\frac{6}{12}\pi\right)}\right)\right).$$

The fifth-degree term of the realized denominator,

$$\omega^5\left(C_2 C_4 C_6 L_3 L_5 R_S + C_2 C_4 L_1 L_3 L_5 \frac{1}{R_L}\right),$$

is associated with the fifth-degree Butterworth term

$$s^5\left(\left(\frac{\cos(0)}{\sin\left(\frac{1}{12}\pi\right)}\right)\left(\frac{\cos\left(\frac{1}{12}\pi\right)}{\sin\left(\frac{2}{12}\pi\right)}\right)\left(\frac{\cos\left(\frac{2}{12}\pi\right)}{\sin\left(\frac{3}{12}\pi\right)}\right)\left(\frac{\cos\left(\frac{3}{12}\pi\right)}{\sin\left(\frac{4}{12}\pi\right)}\right)\left(\frac{\cos\left(\frac{4}{12}\pi\right)}{\sin\left(\frac{5}{12}\pi\right)}\right)\right).$$

The fourth-degree term of the realized denominator,

$$\omega^4\Big(C_2C_4L_1L_3 + C_2C_4L_3L_5\frac{R_S}{R_L} + C_2C_6L_1L_3 + C_2C_6L_1L_5 + C_4C_6L_1L_5 + C_4C_6L_3L_5\Big),$$

is associated with the fourth-degree Butterworth term

$$s^4\left(\left(\frac{\cos(0)}{\sin\left(\frac{1}{12}\pi\right)}\right)\left(\frac{\cos\left(\frac{1}{12}\pi\right)}{\sin\left(\frac{2}{12}\pi\right)}\right)\left(\frac{\cos\left(\frac{2}{12}\pi\right)}{\sin\left(\frac{3}{12}\pi\right)}\right)\left(\frac{\cos\left(\frac{3}{12}\pi\right)}{\sin\left(\frac{4}{12}\pi\right)}\right)\right).$$

The third-degree term of the realized denominator,

$$\omega^3\Big(C_2C_4L_3R_S + C_2C_6L_3R_S + C_2C_6L_5R_S + C_2L_1L_3\frac{1}{R_L} + C_2L_1L_5\frac{1}{R_L} + C_4C_6L_5R_S + C_4L_1L_5\frac{1}{R_L} + C_4L_3L_5\frac{1}{R_L}\Big),$$

is associated with the third-degree Butterworth term

$$s^3\left(\left(\frac{\cos(0)}{\sin\left(\frac{1}{12}\pi\right)}\right)\left(\frac{\cos\left(\frac{1}{12}\pi\right)}{\sin\left(\frac{2}{12}\pi\right)}\right)\left(\frac{\cos\left(\frac{2}{12}\pi\right)}{\sin\left(\frac{3}{12}\pi\right)}\right)\right).$$

The second-degree term of the realized denominator,

$$\omega^2\Big(C_2L_1 + C_2L_3\frac{R_S}{R_L} + C_2L_5\frac{R_S}{R_L} + C_4L_1 + C_4L_3 + C_4L_5\frac{R_S}{R_L} + C_6L_1 + C_6L_3 + C_6L_5\Big),$$

is associated with the second-degree Butterworth term

$$s^2\left(\left(\frac{\cos(0)}{\sin\left(\frac{1}{12}\pi\right)}\right)\left(\frac{\cos\left(\frac{1}{12}\pi\right)}{\sin\left(\frac{2}{12}\pi\right)}\right)\right).$$

The first-degree term of the realized denominator,

$$\omega\left(\frac{L_1}{R_L} + C_2R_S + \frac{L_3}{R_L} + C_4R_S + \frac{L_5}{R_L} + C_6R_S\right),$$

is associated with the first-degree Butterworth term

$$s\left(\left(\frac{\cos(0)}{\sin\left(\frac{1}{12}\pi\right)}\right)\right).$$

Applying the transformation rule yields

$$\begin{aligned}
&AC_{aspect} = (DC_{aspect})\,(Butterworth_{coef})\\
&\omega^6 L_1C_2L_3C_4L_5C_6 = \left(\frac{R_S}{R_L}+1\right)\left(\left(\frac{\cos(0)}{\sin\left(\frac{1}{12}\pi\right)}\right)\left(\frac{\cos(\frac{1}{12}\pi)}{\sin\left(\frac{2}{12}\pi\right)}\right)\right.\\
&\quad\left.\left(\frac{\cos(\frac{2}{12}\pi)}{\sin\left(\frac{3}{12}\pi\right)}\right)\left(\frac{\cos(\frac{3}{12}\pi)}{\sin\left(\frac{4}{12}\pi\right)}\right)\left(\frac{\cos(\frac{4}{12}\pi)}{\sin\left(\frac{5}{12}\pi\right)}\right)\left(\frac{\cos(\frac{5}{12}\pi)}{\sin\left(\frac{6}{12}\pi\right)}\right)\right)\\
&\omega^5\left(C_2C_4C_6L_3L_5R_S + C_2C_4L_1L_3L_5\frac{1}{R_L}\right) = \left(\frac{R_S}{R_L}+1\right)\left(\left(\frac{\cos(0)}{\sin\left(\frac{1}{12}\pi\right)}\right)\left(\frac{\cos(\frac{1}{12}\pi)}{\sin\left(\frac{2}{12}\pi\right)}\right)\right.\\
&\quad\left.\left(\frac{\cos(\frac{2}{12}\pi)}{\sin\left(\frac{3}{12}\pi\right)}\right)\left(\frac{\cos(\frac{3}{12}\pi)}{\sin\left(\frac{4}{12}\pi\right)}\right)\left(\frac{\cos(\frac{4}{12}\pi)}{\sin\left(\frac{5}{12}\pi\right)}\right)\right)\\
&\omega^4\left(C_2C_4L_1L_3 + C_2C_4L_3L_5\frac{R_S}{R_L} + C_2C_6L_1L_3\right.\\
&\quad\left. + C_2C_6L_1L_5 + C_4C_6L_1L_5 + C_4C_6L_3L_5\right)\\
&\quad = \left(\frac{R_S}{R_L}+1\right)\left(\left(\frac{\cos(0)}{\sin\left(\frac{1}{12}\pi\right)}\right)\left(\frac{\cos(\frac{1}{12}\pi)}{\sin\left(\frac{2}{12}\pi\right)}\right)\right.\\
&\quad\left.\left(\frac{\cos(\frac{2}{12}\pi)}{\sin\left(\frac{3}{12}\pi\right)}\right)\left(\frac{\cos(\frac{3}{12}\pi)}{\sin\left(\frac{4}{12}\pi\right)}\right)\right)\\
&\omega^3\left(C_2C_4L_3R_S + C_2C_6L_3R_S + C_2C_6L_5R_S + C_2L_1L_3\frac{1}{R_L}\right.\\
&\quad\left. + C_2L_1L_5\frac{1}{R_L} + C_4C_6L_5R_S + C_4L_1L_5\frac{1}{R_L} + C_4L_3L_5\frac{1}{R_L}\right)\\
&\quad = \left(\frac{R_S}{R_L}+1\right)\left(\left(\frac{\cos(0)}{\sin\left(\frac{1}{12}\pi\right)}\right)\left(\frac{\cos(\frac{1}{12}\pi)}{\sin\left(\frac{2}{12}\pi\right)}\right)\right.\\
&\quad\left.\left(\frac{\cos(\frac{2}{12}\pi)}{\sin\left(\frac{3}{12}\pi\right)}\right)\right)\\
&\omega^2\left(C_2L_1 + C_2L_3\frac{R_S}{R_L} + C_2L_5\frac{R_S}{R_L} + C_4L_1 + C_4L_3\right.\\
&\quad\left. + C_4L_5\frac{R_S}{R_L} + C_6L_1 + C_6L_3 + C_6L_5\right) = \left(\frac{R_S}{R_L}+1\right)\left(\left(\frac{\cos(0)}{\sin\left(\frac{1}{12}\pi\right)}\right)\left(\frac{\cos(\frac{1}{12}\pi)}{\sin\left(\frac{2}{12}\pi\right)}\right)\right)\\
&\omega\left(\frac{L_1}{R_L} + C_2R_S + \frac{L_3}{R_L} + C_4R_S + \frac{L_5}{R_L} + C_6R_S\right) = \left(\frac{R_S}{R_L}+1\right)\left(\left(\frac{\cos(0)}{\sin\left(\frac{1}{12}\pi\right)}\right)\right)
\end{aligned} \tag{11.12}$$

The system of Eq. (11.12) simplifies to

$$\omega^6 L_1 C_2 L_3 C_4 L_5 C_6 = \left(\frac{R_S}{R_L} + 1\right), \tag{11.13}$$

$$\begin{aligned}
\omega^5 \left(C_2 C_4 C_6 L_3 L_5 R_S + C_2 C_4 L_1 L_3 L_5 \frac{1}{R_L}\right) &= \left(2\sqrt{2+\sqrt{3}}\right)\left(\frac{R_S}{R_L} + 1\right) \\
&= \left(\frac{2}{\sqrt{2-\sqrt{3}}}\right)\left(\frac{R_S}{R_L} + 1\right) \\
&= \left(\csc\left(\frac{\pi}{12}\right)\right)\left(\frac{R_S}{R_L} + 1\right),
\end{aligned} \tag{11.14}$$

$$\begin{aligned}
&\omega^4 \Bigg(C_2 C_4 L_1 L_3 + C_2 C_4 L_3 L_5 \frac{R_S}{R_L} + C_2 C_6 L_1 L_3 \\
&\quad + C_2 C_6 L_1 L_5 + C_4 C_6 L_1 L_5 + C_4 C_6 L_3 L_5 \Bigg) \\
&= \left(2\left(2+\sqrt{3}\right)\right)\left(\frac{R_S}{R_L} + 1\right) \\
&= \left(\frac{2}{2-\sqrt{3}}\right)\left(\frac{R_S}{R_L} + 1\right) \\
&= \left(\frac{\csc(\frac{\pi}{12})^2}{2}\right)\left(\frac{R_S}{R_L} + 1\right),
\end{aligned} \tag{11.15}$$

$$\begin{aligned}
&\omega^3 \Bigg(C_2 C_4 L_3 R_S + C_2 C_6 L_3 R_S + C_2 C_6 L_5 R_S + C_2 L_1 L_3 \frac{1}{R_L} + C_2 L_1 L_5 \frac{1}{R_L} \\
&\quad + C_4 C_6 L_5 R_S + C_4 L_1 L_5 \frac{1}{R_L} + C_4 L_3 L_5 \frac{1}{R_L} \Bigg) \\
&= \left(2\left(2+\sqrt{3}\right)\sqrt{\frac{3}{2}}\right)\left(\frac{R_S}{R_L} + 1\right) \\
&= \left(\frac{\csc(\frac{\pi}{12})^2}{2}\sqrt{\frac{3}{2}}\right)\left(\frac{R_S}{R_L} + 1\right) \\
&= \left(\sqrt{6}\left(2+\sqrt{3}\right)\right)\left(\frac{R_S}{R_L} + 1\right),
\end{aligned} \tag{11.16}$$

$$\begin{aligned}
&\omega^2\Big(C_2L_1 + C_2L_3\frac{R_S}{R_L} + C_2L_5\frac{R_S}{R_L} + C_4L_1 + C_4L_3 \\
&\quad + C_4L_5\frac{R_S}{R_L} + C_6L_1 + C_6L_3 + C_6L_5\Big) \\
&= \left(2\left(2+\sqrt{3}\right)\right)\left(\frac{R_S}{R_L}+1\right) \\
&= \left(\frac{2}{2-\sqrt{3}}\right)\left(\frac{R_S}{R_L}+1\right) \\
&= \left(\frac{\csc(\frac{\pi}{12})^2}{2}\right)\left(\frac{R_S}{R_L}+1\right),
\end{aligned} \tag{11.17}$$

and

$$\begin{aligned}
\omega\left(\frac{L_1}{R_L} + C_2R_S + \frac{L_3}{R_L} + C_4R_S + \frac{L_5}{R_L} + C_6R_S\right) &= \left(2\sqrt{2+\sqrt{3}}\right)\left(\frac{R_S}{R_L}+1\right) \\
&= \left(\frac{2}{\sqrt{2-\sqrt{3}}}\right)\left(\frac{R_S}{R_L}+1\right) \\
&= \left(\csc\left(\frac{\pi}{12}\right)\right)\left(\frac{R_S}{R_L}+1\right).
\end{aligned} \tag{11.18}$$

As in the shunt-first case, the sixth-degree term and the constant term are equal, the fifth- and first-degree terms are equal, and the fourth- and second-degree terms are equal. The third-degree term stands alone. This is the sixth-order denominator-level manifestation of the even-order symmetry discussed earlier in Sect. 11.1.

Since there are six equations and six unknown reactive element values, the sixth-order series-first system can produce multiple candidate solution tuples. In practice, the real-valued tuples are retained as exact passive realizations.

11.5 Time Constants

Equations (6.8) and (6.17) define τ_C and τ_L, which are used throughout this section.

For the sixth-order shunt-first low-pass filter, the first-degree portion of the system is

$$\begin{aligned}
&\omega\begin{pmatrix} C_1R_S + \dfrac{L_2}{R_L} + C_3R_S + \dfrac{L_4}{R_L} \\ + C_5R_S + \dfrac{L_6}{R_L} \end{pmatrix} = \left(\frac{R_S}{R_L}+1\right)\left(2\sqrt{2+\sqrt{3}}\right), \\
&\omega\begin{pmatrix} C_1\dfrac{R_SR_L}{R_S+R_L} + \dfrac{L_2}{R_S+R_L} + C_3\dfrac{R_SR_L}{R_S+R_L} \\ + \dfrac{L_4}{R_S+R_L} + C_5\dfrac{R_SR_L}{R_S+R_L} + \dfrac{L_6}{R_S+R_L} \end{pmatrix} = \left(2\sqrt{2+\sqrt{3}}\right).
\end{aligned} \tag{11.19}$$

For the sixth-order series-first low-pass filter, the corresponding first-degree portion is

$$\omega \left(\begin{array}{l} \dfrac{L_1}{R_L} + C_2 R_S + \dfrac{L_3}{R_L} + C_4 R_S \\ + \dfrac{L_5}{R_L} + C_6 R_S \end{array} \right) = \left(\frac{R_S}{R_L} + 1 \right) \left(2\sqrt{2+\sqrt{3}} \right),$$

$$\omega \left(\begin{array}{l} \dfrac{L_1}{R_S + R_L} + C_2 \dfrac{R_S R_L}{R_S + R_L} + \dfrac{L_3}{R_S + R_L} \\ + C_4 \dfrac{R_S R_L}{R_S + R_L} + \dfrac{L_5}{R_S + R_L} + C_6 \dfrac{R_S R_L}{R_S + R_L} \end{array} \right) = \left(2\sqrt{2+\sqrt{3}} \right). \tag{11.20}$$

Thus, both sixth-order topologies resolve to

$$\begin{aligned} \omega \left(\tau_{C_1} + \tau_{L_2} + \tau_{C_3} + \tau_{L_4} + \tau_{C_5} + \tau_{L_6} \right) &= \omega \left(\tau_{L_1} + \tau_{C_2} + \tau_{L_3} + \tau_{C_4} + \tau_{L_5} + \tau_{C_6} \right) \\ &= \frac{1}{\sin(\frac{1}{12}\pi)}. \end{aligned}$$

Equivalently,

$$\begin{aligned} &\omega\tau_{C_1} + \omega\tau_{L_2} + \omega\tau_{C_3} + \omega\tau_{L_4} + \omega\tau_{C_5} + \omega\tau_{L_6} \\ &= \omega\tau_{L_1} + \omega\tau_{C_2} + \omega\tau_{L_3} + \omega\tau_{C_4} + \omega\tau_{L_5} + \omega\tau_{C_6} = \frac{1}{\sin(\frac{1}{12}\pi)}. \end{aligned} \tag{11.21}$$

For equal terminations, the individual normalized time constants are not generally equal to the finite sine terms one-by-one for every real-valued solution tuple. However, the summed identity remains valid:

$$\begin{aligned} &\sin\left(\frac{1}{12}\pi\right) + \sin\left(\frac{3}{12}\pi\right) + \sin\left(\frac{5}{12}\pi\right) \\ &+ \sin\left(\frac{7}{12}\pi\right) + \sin\left(\frac{9}{12}\pi\right) + \sin\left(\frac{11}{12}\pi\right) = \frac{1}{\sin(\frac{1}{12}\pi)}. \end{aligned} \tag{11.22}$$

This is the sixth-order instance of the finite trigonometric time-constant structure that is later collected in Chap. 12.

11.6 High-Pass Shunt-First

A sixth-order shunt-first high-pass filter is obtained by first synthesizing the corresponding shunt-first low-pass ladder and then converting each reactive element using Eqs. (6.19) and (6.20). Therefore, the high-pass case is not a separate coefficient-matching derivation.

It is a transformation example that uses the corresponding sixth-order low-pass system to obtain a low-pass precursor, which is then converted into the final high-pass ladder. For $R_S > R_L$, this is the appropriate sixth-order high-pass topology (Fig. 11.3).

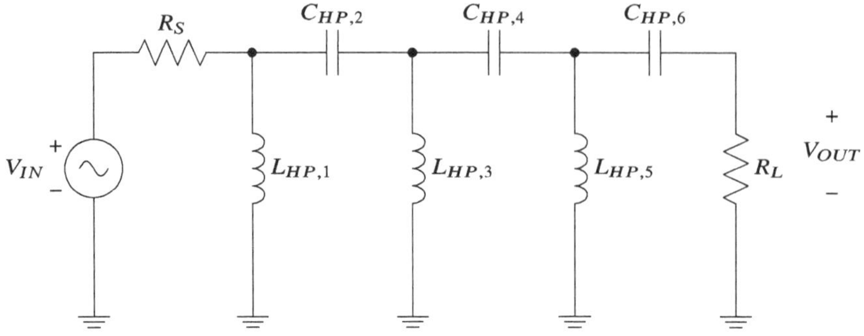

Fig. 11.3 Sixth-order shunt-first high-pass Butterworth filter

For a shunt-first low-pass realization, the corresponding high-pass element values are

$$
\begin{aligned}
L_{HP,1} &= \frac{1}{\omega_c^2 C_1}, & C_{HP,2} &= \frac{1}{\omega_c^2 L_2}, \\
L_{HP,3} &= \frac{1}{\omega_c^2 C_3}, & C_{HP,4} &= \frac{1}{\omega_c^2 L_4}, \\
L_{HP,5} &= \frac{1}{\omega_c^2 C_5}, & C_{HP,6} &= \frac{1}{\omega_c^2 L_6}.
\end{aligned}
$$

11.7 High-Pass Series-First

A sixth-order series-first high-pass filter is obtained by first synthesizing the corresponding sixth-order series-first low-pass ladder and then converting each reactive element using Eqs. (6.19) and (6.20). Therefore, the high-pass case is not a separate coefficient-matching derivation.

It is a transformation example that uses the corresponding sixth-order low-pass system to obtain a low-pass precursor, which is then converted into the final high-pass ladder. For $R_S < R_L$, this is the complementary sixth-order high-pass topology (Fig. 11.4).

For a series-first low-pass realization, the corresponding high-pass element values are

$$
\begin{aligned}
C_{HP,1} &= \frac{1}{\omega_c^2 L_1}, & L_{HP,2} &= \frac{1}{\omega_c^2 C_2}, \\
C_{HP,3} &= \frac{1}{\omega_c^2 L_3}, & L_{HP,4} &= \frac{1}{\omega_c^2 C_4}, \\
C_{HP,5} &= \frac{1}{\omega_c^2 L_5}, & L_{HP,6} &= \frac{1}{\omega_c^2 C_6}.
\end{aligned}
$$

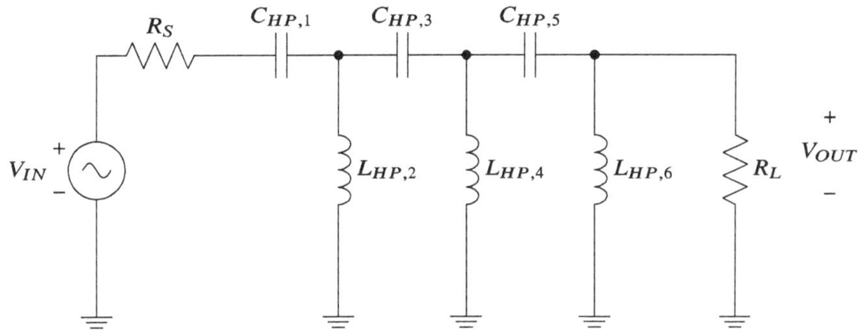

Fig. 11.4 Sixth-order series-first high-pass Butterworth filter

11.8 Conclusion

The sixth-order case extends the exact synthesis method without changing its principle. The denominator is still matched degree by degree to the corresponding Butterworth structure, but the pattern of symmetry now takes its higher even-order form: sixth degree paired with the constant term, fifth degree paired with first degree, fourth degree paired with second degree, and third degree standing alone.

This chapter also makes visible the factor-family transition introduced in Chap. 10. In fifth order, the secondary correlated factor family evaluates to unity and therefore does not visibly alter the normalized fifth-order coefficient structure. In sixth order, the same structural feature appears as the reciprocal pair $\sqrt{3/2}$ and $1/\sqrt{3/2}$, and therefore must be retained explicitly in the product-factor interpretation.

Together, Chaps. 9 and 10, and this chapter show the before-transition-afterward sequence: fourth order retains the primary factor-family structure, fifth order introduces the secondary correlated factor family in transparent form, and sixth order makes that secondary family visible in the polynomial coefficients. This also marks the point at which exhaustive numerical design examples begin to produce diminishing returns. The low-pass and high-pass synthesis structures are retained, while the repeated numerical design examples are intentionally omitted so that the sixth-order factor-family structure remains the focus.

References

1. Butterworth, S. (1930). On the theory of filter amplifiers. *Experimental Wireless and the Wireless Engineer, 7*, 536–541.
2. Zverev, A. I. (1967). *Handbook of filter synthesis*. Wiley.

Part III
Mathematical Extensions

Abstract This chapter collects several finite trigonometric series, product identities, and factor-level relationships that emerge from the Butterworth synthesis framework developed in the preceding chapters. In particular, the recurring time-constant relationships observed in first through fourth-order doubly-terminated filters are expressed in compact cosecant and secant summation formulas, together with a related product identity involving inverse powers of two. The chapter then examines the internal structure of Butterworth product factors, showing how lower-order factors exhibit simple mutual relationships while higher-order cases reintroduce more explicit recursive structure. Finally, the chapter presents element-product formulas that may be used to analyze the cutoff frequency of existing doubly-terminated Butterworth filters.

12 Mathematical Identities Emerging from Butterworth Synthesis

Patterns that recur are asking to be named.

Chapters 6 through 9 applied the same synthesis method to filters of increasing order. Although the number of reactive elements and the degree of the circuit denominator increased with filter order, the underlying transformation remained the same. Repeated application of that transformation produced the time-constant sequences, Butterworth factor relationships, and denominator structures from which the mathematical patterns examined in this chapter became visible. The purpose of the present chapter is to collect those recurring patterns into explicit identities and to place them in their appropriate mathematical context.

Before examining those patterns individually, it is useful to reconsider the transformation that generated them:

$$AC_{\text{aspect}} = DC_{\text{aspect}} \times Butterworth_{\text{coefficient}}.$$

In the order-specific synthesis chapters, this relationship was used constructively to determine the reactive-element values required for Butterworth synthesis. Viewed retrospectively, however, it reveals a more fundamental feature of the method: the synthesis does not treat gain and frequency as separate stages. The DC_{aspect} establishes the realized passband gain and therefore supplies the vertical reference from which the Butterworth cutoff level is measured. The reactive terms contained in the AC_{aspect} supply the horizontal frequency dependence. The Butterworth coefficients provide the multiplicative relationship that ties

K. M. P. Ryan, *Polynomial Symmetry in Butterworth Filter Synthesis*,
Synthesis Lectures on Engineering, Science, and Technology,
https://doi.org/10.1007/978-3-032-33468-8_12

these two dimensions together. Through the Butterworth coefficients, the corresponding AC_{aspect} terms are constrained relative to the common DC_{aspect} reference throughout the denominator.

No single polynomial degree establishes the cutoff condition independently; rather, the complete denominator must satisfy the Butterworth coefficient structure collectively. When these relationships are satisfied across the full denominator, the realized response reaches the exact half-power Butterworth cutoff condition, corresponding to 3.0103 dB below the DC_{aspect} reference, at the specified cutoff frequency. The gain datum is global, the frequency dependence is distributed through the reactive terms, and the exact cutoff emerges from their collective Butterworth relationship across the complete denominator.

It is through repeated application of this same transformation that the additional mathematical structure collected in this chapter emerged. As filter order increased, recurring time-constant sums became visible, product relationships appeared among the Butterworth factors, and additional relationships arose among the reactive-element products. The recurring time-constant relationships provide the immediate starting point for the three finite trigonometric identities considered below.

The three finite relationships presented in Sects. 12.1, 12.2 and 12.3 were identified independently through the Butterworth calculations themselves rather than introduced from pre-existing trigonometric formulas. The underlying finite trigonometric identities are classical, and the odd-angle sine summation in Eq. (12.1) has also appeared in prior filter literature in a prototype-coefficient setting [1]. Standard finite trigonometric sum and product formulas provide independent verification of the relationships [2]. The contribution claimed here is therefore not mathematical priority for the underlying identities, but the identification of their correspondence with the direct doubly-terminated synthesis and the recurring time-constant structure developed in the preceding chapters.

This chapter therefore serves as a mathematical culmination of the synthesis developed in this monograph. It does not introduce a new synthesis method. Instead, it extracts structural consequences of the method already developed: the finite trigonometric identities that govern the summed time constants, the product identities that accompany them, the factor-level symmetries that help explain lower-order transparency, and the element-product formulas that permit cutoff analysis of existing filters.

12.1 A Finite Summation Identity for Cosecant

Based on Eqs. (6.9), (7.16), (8.16) and (9.18), the following sequence of identities appears:

$$\sin\left(\frac{1}{2}\pi\right) = \frac{1}{\sin\left(\frac{1}{2}\pi\right)} = \csc\left(\frac{1}{2}\pi\right),$$

$$\sin\left(\frac{1}{4}\pi\right) + \sin\left(\frac{3}{4}\pi\right) = \frac{1}{\sin\left(\frac{1}{4}\pi\right)} = \csc\left(\frac{1}{4}\pi\right),$$

$$\sin\left(\frac{1}{6}\pi\right) + \sin\left(\frac{3}{6}\pi\right) + \sin\left(\frac{5}{6}\pi\right) = \frac{1}{\sin\left(\frac{1}{6}\pi\right)} = \csc\left(\frac{1}{6}\pi\right),$$

$$\sin\left(\frac{1}{8}\pi\right) + \sin\left(\frac{3}{8}\pi\right) + \sin\left(\frac{5}{8}\pi\right) + \sin\left(\frac{7}{8}\pi\right) = \frac{1}{\sin\left(\frac{1}{8}\pi\right)} = \csc\left(\frac{1}{8}\pi\right).$$

These examples motivate the finite summation identity

$$\sum_{i=1}^{n} \sin\left((2i-1)\frac{\pi}{2n}\right) = \frac{1}{\sin\left(\frac{\pi}{2n}\right)} = \csc\left(\frac{\pi}{2n}\right). \quad (12.1)$$

Equation (12.1) is the compact form of the time-constant summation pattern that first appears in the order-specific synthesis chapters. The identity itself is classical and follows from the standard finite sum of sines in an arithmetic progression [2]. An equivalent odd-angle sine summation has also appeared in connection with maximally-flat low-pass prototype coefficients [1]. Its role here is different: the same angular structure was identified independently through the time-constant relationships produced by the direct doubly-terminated synthesis.

12.2 A Finite Summation Identity for Secant

A related identity may be written for secant. Since secant is offset from cosecant by $\pi/2$, the corresponding finite cosine sum is

$$\sum_{i=1}^{n} \cos\left((2i-n-1)\frac{\pi}{2n}\right) = \frac{1}{\cos\left((n-1)\frac{\pi}{2n}\right)} = \sec\left((n-1)\frac{\pi}{2n}\right). \quad (12.2)$$

This corresponding secant form was likewise identified from the recurring Butterworth structure. It is the cofunction counterpart of Eq. (12.1) and may be independently verified using the standard finite sum of cosines in an arithmetic progression [2].

Together, Eqs. (12.1) and (12.2) show that the trigonometric structure emerging from Butterworth synthesis is not confined to a single representation. The same order-dependent structure may be viewed through either sine-based or cosine-based identities.

12.3 A Product Identity for Inverse Powers of Two

A second recurring pattern appears in the product of the same odd-indexed angular samples. In compact form,

$$\prod_{i=1}^{n} \sin\left((2i-1)\frac{\pi}{2n}\right) = \prod_{i=1}^{n} \cos\left((2i-1-n)\frac{\pi}{2n}\right) = \frac{1}{2^{(n-1)}}. \tag{12.3}$$

This inverse-power-of-two relationship was identified from the recurring Butterworth factor calculations. The underlying identity is classical and may be independently verified using the standard finite sine-product formula [2].

This identity is useful in its own right, but it is also suggestive of the deeper factor structure underlying the Butterworth coefficients. The finite product does not merely produce a number; it reflects how the Butterworth factors are distributed about the center.

12.4 Butterworth Product Factors as Structural Building Blocks

Butterworth polynomial coefficients are commonly generated recursively from product formulas such as Eq. (2.1). Those coefficients, however, are themselves composed of lower-level Butterworth product factors. In that sense, the factors may be regarded as the elementary building blocks from which the Butterworth coefficients are constructed.

This perspective matters because the Butterworth character of the polynomial is carried not only by the final coefficient list, but also by the structured relationships among the factors from which those coefficients are formed. The factors are therefore the smallest recurring objects in the present development.

For lower orders, these factor relationships are especially transparent. In orders 1 through 4, the Butterworth product factors can often be expressed in terms of one another through simple scaling, inversion, or pairing relationships. Beginning with fifth order and continuing upward, this direct mutual expressibility becomes less complete, and recursive structure re-emerges more strongly.

12.5 Lower-Order Factor Symmetries

12.5.1 Third Order

For the third-order Butterworth polynomial in the product-factor form used in this monograph, the product factors are 2, 1, and 0.5. Their mutual relationships may be written as

$$
\begin{aligned}
\boxed{2} &= 2(1) = \frac{1}{0.5}, \\
\boxed{1} &= \frac{2}{2} = \frac{1}{2(0.5)}, \\
\boxed{0.5} &= \frac{1}{2} = \frac{1}{2(1)}.
\end{aligned}
\tag{12.4}
$$

They also satisfy the multiplicative relationships

$$
\begin{aligned}
(2)(0.5) &= 1, \\
(2)(1) &= 2, \\
\frac{1}{0.5} &= 2.
\end{aligned}
\tag{12.5}
$$

These relations show that, at third order, all Butterworth product factors may be expressed in terms of each other through elementary operations. In that sense, the third-order case exhibits a particularly transparent factor symmetry.

12.5.2 Fourth Order

For the fourth-order Butterworth polynomial in the product-factor form used in this monograph, the product factors are approximately

$$
2.613125, \quad 1.306562, \quad 0.7653668, \quad 0.3826834.
$$

Their individual relationships may be written as

$$
\begin{aligned}
\boxed{2.613125} &= 2(1.306562) = 2\left(\frac{1}{0.7653668}\right) = \frac{1}{0.3826834}, \\
\boxed{1.306562} &= \frac{2.613125}{2} = \frac{1}{0.7653668} = \frac{1}{2(0.3826834)}, \\
\boxed{0.7653668} &= \frac{2}{2.613125} = \frac{1}{1.306562} = 2(0.3826834), \\
\boxed{0.3826834} &= \frac{1}{2.613125} = \frac{1}{2(1.306562)} = \frac{0.7653668}{2}.
\end{aligned}
\tag{12.6}
$$

Their multiplicative relationships include

$$
\begin{aligned}
(2.613125)(0.3826834) &= 1, \\
(2.613125)(0.7653668) &= 2, \\
(2.613125)(1.306562) &= 2+\sqrt{2}, \\
(1.306562)(0.7653668) &= 1, \\
(1.306562)(0.3826834) &= \frac{1}{2}, \\
(0.7653668)(0.3826834) &= \frac{1}{2+\sqrt{2}}, \\
(2.613125)^2 &= 2(2+\sqrt{2}).
\end{aligned}
\tag{12.7}
$$

At fourth order, the factor structure is still highly interrelated. The factors may be expressed through one another by inversion, halving, doubling, and pairing. This helps explain why the lower-order synthesis chapters retain a relatively high degree of analytical transparency.

12.6 The Onset of Recursive Structure

12.6.1 Fifth Order

Beginning with fifth order, the mutual expressibility of the product factors becomes less complete. The factor relationships remain rich, but the internal structure is no longer exhausted by simple inversion and scaling.

The five principal factors may be summarized by the representative values

$$
3.236067, \quad 1.6180339, \quad 1, \quad 0.6180339, \quad 0.3090169.
$$

Several relationships remain simple:

$$
\begin{aligned}
\boxed{3.236067} &= 2(1.6180339) = 2\left(\frac{1}{0.6180339}\right) = \frac{1}{0.3090169}, \\
\boxed{1.6180339} &= \frac{3.236067}{2} = \frac{1}{0.6180339} = \frac{1}{2(0.3090169)}, \\
\boxed{0.6180339} &= \frac{2}{3.236067} = \frac{1}{1.6180339} = 2(0.3090169), \\
\boxed{0.3090169} &= \frac{1}{3.236067} = \frac{1}{2(1.6180339)} = \frac{0.6180339}{2}.
\end{aligned}
\tag{12.8}
$$

But the middle factor now stands apart:

$$
1.
$$

This already signals a transition. The fifth-order case still retains a high degree of internal structure, and in this specific case the middle factor is especially simple, but the factor system is no longer fully reducible to one generating relation in the same way as the lower orders.

12.6.2 Sixth Order

By sixth order, the recurrence structure is more explicit. The outer and near-outer factors remain highly interrelated, but the middle factors become only partially linked to the rest of the system. In the notation of the sixth-order factor table, the factors on Lines 1, 2, 5, and 6 remain mutually expressible through simple scaling and inversion relationships, while the factors on Lines 3 and 4 are related primarily through inversion of each other rather than through the full outer pattern.

This is the point at which recursion becomes more than a technical convenience. It becomes part of the structure itself. The factor pattern is still organized, but no longer fully collapsible into the simpler lower-order relationships.

A concise way to state the progression is:

> For Butterworth orders 1 through 4, the product factors exhibit a high degree of mutual expressibility through simple scaling and inversion relationships. Beginning with fifth order, this simplicity becomes incomplete, and recursive structure reappears more explicitly.

12.7 Element-Product Formulas for Existing Filters

The element-product formulas developed in this section provide a direct means of determining the cutoff frequency of an already designed doubly-terminated Butterworth filter from its known reactive-element values and source/load terminations. Unlike the complete synthesis procedure, this analysis does not require reconstruction of every polynomial degree.

For an nth-order network, the required relationship is contained in the highest-degree term of the circuit denominator. The highest-degree Butterworth coefficient is unity, while the corresponding highest-degree circuit term contains the multiplicative product of all n reactive elements together with the topology-dependent source/load resistance factor. The highest-degree coefficient-matching equation may therefore be isolated and rearranged for ω^n, allowing the cutoff frequency to be recovered directly from the existing component values and terminations. No lower-degree polynomial equations are required for this calculation.

This reduction applies because the network under examination is already a Butterworth realization. It should not be interpreted to mean that the highest-degree term alone establishes the Butterworth response during synthesis. The exact cutoff behavior of the designed filter remains a property of the complete denominator; the highest-degree equation merely contains sufficient information to recover the cutoff frequency once that Butterworth structure has already been established.

In the compact notation used below, the product symbols denote multiplicative products of the topology-specific alternating reactive elements. No division between inductive and capacitive element products is implied.

Taking only the highest-degree relationship from Eqs. (6.5), (6.14), (7.4), (7.11), (8.4), (8.10), (9.4), and (9.11) yields the following four topology-dependent element-product relationships.

12.7.1 Even-Order Shunt-First Filters for $R_S > R_L$

$$\left(\prod_{k=1}^{n/2} C_{2k-1}\right)\left(\prod_{k=1}^{n/2} L_{2k}\right) = \frac{1}{\omega^n \left(\frac{R_S}{R_S+R_L}\right)}. \tag{12.9}$$

12.7.2 Even-Order Series-First Filters for $R_S < R_L$

$$\left(\prod_{k=1}^{n/2} L_{2k-1}\right)\left(\prod_{k=1}^{n/2} C_{2k}\right) = \frac{1}{\omega^n \left(\frac{R_L}{R_S+R_L}\right)}. \tag{12.10}$$

12.7.3 Odd-Order Shunt-First Filters

$$\left(\prod_{k=1}^{(n+1)/2} C_{2k-1}\right)\left(\prod_{k=1}^{(n-1)/2} L_{2k}\right) = \frac{1}{\omega^n \left(\frac{R_S R_L}{R_S+R_L}\right)}. \tag{12.11}$$

12.7.4 Odd-Order Series-First Filters

$$\left(\prod_{k=1}^{(n+1)/2} L_{2k-1}\right)\left(\prod_{k=1}^{(n-1)/2} C_{2k}\right) = \frac{1}{\omega^n \left(\frac{1}{R_S+R_L}\right)}. \tag{12.12}$$

These formulas are useful because they permit cutoff analysis of an existing passive doubly-terminated Butterworth ladder filter without reconstructing the full synthesis step by step. Once the product of the relevant reactive elements is known, ω^n may be isolated and the cutoff frequency recovered directly.

12.8 Conclusion

This chapter gathers the mathematical patterns that emerged gradually in the synthesis chapters and presents them as explicit identities. The finite trigonometric sums explain the recurring time-constant relations. The product identity for inverse powers of two reveals a complementary order-dependent structure. The factor-level analysis clarifies why lower-order Butterworth synthesis remains especially transparent and why higher-order cases reintroduce more explicit recursion.

The interpretation developed at the beginning of this chapter also makes explicit what the synthesis transformation has been doing throughout the order-specific designs. The DC_{aspect} supplies the vertical gain datum, the reactive terms contained in the AC_{aspect} supply the horizontal frequency dependence, and the Butterworth coefficients tie the two together multiplicatively. The resulting synthesis condition is therefore simultaneously a gain condition and a frequency condition, with the Butterworth cutoff occurring approximately 3 dB below the realized DC level at the specified cutoff frequency.

The finite cosecant, secant, and inverse-power-of-two relationships were identified through repeated application of that synthesis. Their underlying trigonometric forms are established mathematical results [2], and the cosecant summation has prior use in filter-coefficient analysis [1]. Their significance here is the circuit-synthesis correspondence: the relationships emerged from the time-constant and factor structure of the direct doubly-terminated design rather than serving as the starting point for it.

In this way, the chapter serves as more than a collection of formulas. It shows that the synthesis framework developed in this monograph is not only constructive, but also generative: once the method is applied repeatedly, additional mathematical structure begins to reveal itself.

References

1. Saeedi, S., Lee, J., & Sigmarsson, H. H. (2016). A new property of maximally-flat lowpass filter prototype coefficients with application in dissipative loss calculations. *Progress in Electromagnetics Research C, 63*, 1–11. https://doi.org/10.2528/PIERC16011304
2. Gradshteyn, I. S., & Ryzhik, I. M. (2007). *Table of integrals, series, and products* (7th ed.). Academic Press.

Butterworth Polynomial Factor Relationships

A

A.1 Symmetry of Butterworth Polynomial Factors

Butterworth Polynomial Factors have been generally produced using recursive equations, such as Eq. (2.1). Once the Butterworth Polynomial Factors have been produced, or as they are being produced, they are multiplied together, in order, to produce Butterworth Polynomial Coefficients.

Butterworth Polynomial Factors may be thought of as sub-parts or building blocks of Butterworth Polynomial Coefficients.

Butterworth Polynomial Coefficients, which are made up of Butterworth Polynomial Factors, are what give a polynomial its unique Butterworth, or monotonic, quality. Therefore, the Butterworth Polynomial Factors are the smallest common denominator in determining the monotonic characteristic.

The interrelationships among Butterworth Polynomial Factors are important in considering monotonicity.

There is another unique aspect of Butterworth Polynomial Factors, other than being subparts of Butterworth Polynomial Coefficients, and that is that they may be expressed in terms of each other for filter orders 1–4. In filter orders 1–4, all Butterworth Polynomial Factors may be expressed in terms of each other. This alleviates the need for recursive mathematical operations for filter orders 1–4. However, beginning with the 5th-order Butterworth Filter and higher, the need for recursive mathematics resurfaces.

K. M. P. Ryan, *Polynomial Symmetry in Butterworth Filter Synthesis*,
Synthesis Lectures on Engineering, Science, and Technology,
https://doi.org/10.1007/978-3-032-33468-8

A.1.1 3rd Order

Referencing the product formula for Butterworth polynomial factors, Eq. (2.1), the factors for the normalized third-order Butterworth polynomial are 2, 1, and 0.5.

The relationships among these factors lead to the variations in Table A.1.

Variations of 3rd-Order Butterworth Factors, highlighted in boxes, are presented with respect to each other:

$$\begin{gathered} \boxed{2} = 2(1) = \left(\frac{1}{0.5}\right) \\ \frac{2}{2} = \boxed{1} = \frac{1}{2(0.5)} \\ \frac{1}{2} = \frac{1}{2(1)} = \boxed{0.5} \end{gathered} \tag{A.1}$$

Variations of 3rd-Order Butterworth Factors multiplied with each other:

$$\begin{aligned} (2)(0.5) &= 1 \\ (2)(1) &= 2 \\ \frac{1}{0.5} &= 2 \end{aligned} \tag{A.2}$$

Table A.1 Third-order Butterworth polynomial factors with respect to Eq. (2.1) (variations 1–3), and derivation (variation 4)

Line	Polynomial factor	Variation 1	Variation 2	Variation 3	Variation 4
1	2	$\left(\frac{\cos(0)}{\sin\left(\frac{1}{6}\pi\right)}\right)$	$2\left(\frac{\cos\left(\frac{1}{6}\pi\right)}{\sin\left(\frac{2}{6}\pi\right)}\right)$	$\frac{1}{\left(\frac{\cos\left(\frac{3}{6}\pi\right)}{\sin\left(\frac{4}{6}\pi\right)}\right)}$	$\csc\left(\frac{\pi}{6}\right)$
2	1	$\frac{\left(\frac{\cos(0)}{\sin\left(\frac{1}{6}\pi\right)}\right)}{2}$	$\left(\frac{\cos\left(\frac{1}{6}\pi\right)}{\sin\left(\frac{2}{6}\pi\right)}\right)$	$\frac{1}{2\left(\frac{\cos\left(\frac{3}{6}\pi\right)}{\sin\left(\frac{4}{6}\pi\right)}\right)}$	$\frac{\csc\left(\frac{\pi}{6}\right)}{2}$
3	0.5	$\frac{2}{\left(\frac{\cos(0)}{\sin\left(\frac{1}{6}\pi\right)}\right)}$	$\frac{1}{\left(\frac{\cos\left(\frac{1}{6}\pi\right)}{\sin\left(\frac{2}{6}\pi\right)}\right)}$	$2\left(\frac{\cos\left(\frac{3}{6}\pi\right)}{\sin\left(\frac{4}{6}\pi\right)}\right)$	$\frac{2}{\csc\left(\frac{\pi}{6}\right)}$

In Line 1, the first polynomial factor, 2, may be expressed in terms of itself $\left(\frac{\cos(0)}{\sin\left(\frac{1}{6}\pi\right)}\right)$, the second, and the third polynomial factors.

It is important to point out that all Butterworth polynomial factors may be defined by a single variation. For example, only $\left(\frac{\cos(0)}{\sin\left(\frac{1}{6}\pi\right)}\right)$ need to be calculated. The result of this calculation may be used to define the other two Butterworth Polynomial Factors, 1 (by dividing by 2) and 0.5 (inverting and multiplying by 2).

A.1.2 4th Order

Referencing the product formula for Butterworth polynomial factors, Eq. (2.1), the factors for the normalized fourth-order Butterworth polynomial are 2.613125, 1.306562, 0.7653668, and 0.3826834.

The relationships among these factors are interesting and lead to the variations in Table A.2.

Variations of 4th-Order Butterworth Factors, highlighted in boxes, are presented with respect to each other:

$$\begin{gathered}
\boxed{2.613125} = 2(1.306562) = 2\left(\frac{1}{0.7653668}\right) = \left(\frac{1}{0.3826834}\right) \\
\frac{2.613125}{2} = \boxed{1.306562} = \frac{1}{0.7653668} = \frac{1}{2(0.3826834)} \\
\frac{2}{2.613125} = \frac{1}{1.306562} = \boxed{0.7653668} = 2(0.3826834) \\
\frac{1}{2.613125} = \frac{1}{2(1.306562)} = \frac{0.7653668}{2} = \boxed{0.3826834}
\end{gathered} \tag{A.3}$$

Variations of 4th-Order Butterworth Factors multiplied with each other:

Table A.2 Fourth-order Butterworth polynomial factors with respect to Eq. (2.1) (variations 1–4), and derivations (variations 5–7)

Line	Factor	Variation 1	Variation 2	Variation 3	Variation 4	Variation 5	Variation 6	Variation 7
1	2.613125	$\left(\frac{\cos(0)}{\sin\left(\frac{1}{8}\pi\right)}\right)$	$2\left(\frac{\cos\left(\frac{1}{8}\pi\right)}{\sin\left(\frac{2}{8}\pi\right)}\right)$	$\frac{2}{\left(\frac{\cos\left(\frac{2}{8}\pi\right)}{\sin\left(\frac{3}{8}\pi\right)}\right)}$	$\frac{1}{\left(\frac{\cos\left(\frac{3}{8}\pi\right)}{\sin\left(\frac{4}{8}\pi\right)}\right)}$	$\left(\sqrt{2}\left(\sqrt{2+\sqrt{2}}\right)\right)$	$\left(\frac{2}{\left(\sqrt{2-\sqrt{2}}\right)}\right)$	$\csc\left(\frac{\pi}{8}\right)$
2	1.306562	$\frac{\left(\frac{\cos(0)}{\sin\left(\frac{1}{8}\pi\right)}\right)}{2}$	$\left(\frac{\cos\left(\frac{1}{8}\pi\right)}{\sin\left(\frac{2}{8}\pi\right)}\right)$	$\frac{1}{\left(\frac{\cos\left(\frac{2}{8}\pi\right)}{\sin\left(\frac{3}{8}\pi\right)}\right)}$	$\frac{1}{2\left(\frac{\cos\left(\frac{3}{8}\pi\right)}{\sin\left(\frac{4}{8}\pi\right)}\right)}$	$\frac{\left(\sqrt{2}\left(\sqrt{2+\sqrt{2}}\right)\right)}{2}$	$\left(\frac{1}{\left(\sqrt{2-\sqrt{2}}\right)}\right)$	$\frac{\csc\left(\frac{\pi}{8}\right)}{2}$
3	0.7653668	$\frac{2}{\left(\frac{\cos(0)}{\sin\left(\frac{1}{8}\pi\right)}\right)}$	$\frac{1}{\left(\frac{\cos\left(\frac{1}{8}\pi\right)}{\sin\left(\frac{2}{8}\pi\right)}\right)}$	$\left(\frac{\cos\left(\frac{2}{8}\pi\right)}{\sin\left(\frac{3}{8}\pi\right)}\right)$	$2\left(\frac{\cos\left(\frac{3}{8}\pi\right)}{\sin\left(\frac{4}{8}\pi\right)}\right)$	$\frac{2}{\left(\sqrt{2}\left(\sqrt{2+\sqrt{2}}\right)\right)}$	$\left(\left(\sqrt{2-\sqrt{2}}\right)\right)$	$\frac{2}{\csc\left(\frac{\pi}{8}\right)}$
4	0.3826834	$\frac{1}{\left(\frac{\cos(0)}{\sin\left(\frac{1}{8}\pi\right)}\right)}$	$\frac{1}{2\left(\frac{\cos\left(\frac{1}{8}\pi\right)}{\sin\left(\frac{2}{8}\pi\right)}\right)}$	$\frac{\left(\frac{\cos\left(\frac{2}{8}\pi\right)}{\sin\left(\frac{3}{8}\pi\right)}\right)}{2}$	$\left(\frac{\cos\left(\frac{3}{8}\pi\right)}{\sin\left(\frac{4}{8}\pi\right)}\right)$	$\frac{1}{\left(\sqrt{2}\left(\sqrt{2+\sqrt{2}}\right)\right)}$	$\frac{\left(\left(\sqrt{2-\sqrt{2}}\right)\right)}{2}$	$\frac{1}{\csc\left(\frac{\pi}{8}\right)}$

$$
\begin{aligned}
(2.613125)(0.3826834) &= 1 \\
(2.613125)(0.7653668) &= 2 \\
(2.613125)(1.306562) &= (2+\sqrt{2}) = \frac{2}{(2-\sqrt{2})} \\
(1.306562)(0.7653668) &= 1 \\
(1.306562)(0.3826834) &= \frac{1}{2} \\
(0.7653668)(0.3826834) &= \frac{1}{(2+\sqrt{2})} = \frac{(2-\sqrt{2})}{2} \\
(2.613125)^2 &= 2(2+\sqrt{2}) = \frac{4}{(2-\sqrt{2})}
\end{aligned}
\tag{A.4}
$$

The factors 2.613125, 1.306562, 0.7653668, and 0.3826834 multiply cumulatively to form the normalized fourth-order Butterworth polynomial as:

$$
\begin{aligned}
&s^4 + (\text{line } 1)s^3 + (\text{line } 1 * \text{line } 2)s^2 + (\text{line } 1 * \text{line } 2 * \text{line } 3)s \\
&\quad + (\text{line } 1 * \text{line } 2 * \text{line } 3 * \text{line } 4)
\end{aligned}
$$

which simplifies to

$$
s^4 + (\text{line } 1)s^3 + (\text{line } 1 * \text{line } 2)s^2 + (\text{line } 1)s + 1
$$

A.1.3 5th Order

Variations of 5th-Order Butterworth Factors, highlighted in boxes, are presented with respect to each other:

$$
\begin{aligned}
&\boxed{3.236067} = 2(1.6180339) = 2\left(\frac{1}{0.6180339}\right) = \left(\frac{1}{0.3090169}\right) \\
&\frac{3.236067}{2} = \boxed{1.6180339} = \frac{1}{0.6180339} = \frac{1}{2(0.3090169)} \\
&\frac{2}{3.236067} = \frac{1}{1.6180339} = \boxed{0.6180339} = 2(0.3090169) \\
&\frac{1}{3.236067} = \frac{1}{2(1.6180339)} = \frac{0.6180339}{2} = \boxed{0.3090169}
\end{aligned}
\tag{A.5}
$$

Variations of 5th-Order Butterworth Factors are presented as multiples of each other:

$$\begin{aligned}
(3.236067)(0.3090169) &= 1 \\
(3.236067)(0.6180339) &= 2 \\
(3.236067)(1.6180339) &= 2(1+\Phi) \\
(1.6180339)(0.6180339) &= 1 \\
(1.6180339)(0.3090169) &= \frac{1}{2} \\
(0.6180339)(0.3090169) &= \frac{1}{2(1+\Phi)} \\
(3.236067)^2 &= 4(1+\Phi)
\end{aligned} \tag{A.6}$$

Starting with the 5th-order case, Table A.3, the need for recursive factor multiplication begins to surface. Line 3 marks the first appearance of a secondary correlated factor family. In the 5th-order case this factor evaluates to unity, so it does not alter the simplified polynomial coefficients. However, its appearance is structural: it shows that the complete factor pattern is no longer exhausted by the primary correlated family appearing in the outer lines.

Using Variation 1, or subsequently Variation 8, which is directly related to Variation 1, from Table A.3, we can form the normalized 5th-order Butterworth Polynomial as:

$$\begin{aligned}
&s^5 + (\text{line } 1)s^4 + (\text{line } 1 * \text{line } 2)s^3 \\
&\quad + (\text{line } 1 * \text{line } 2 * \text{line } 3)s^2 \\
&\quad + (\text{line } 1 * \text{line } 2 * \text{line } 3 * \text{line } 4)s \\
&\quad + (\text{line } 1 * \text{line } 2 * \text{line } 3 * \text{line } 4 * \text{line } 5)
\end{aligned}$$

which simplifies to:

$$\begin{aligned}
&s^5 + (\text{line } 1)s^4 + (\text{line } 1 * \text{line } 2)s^3 \\
&\quad + (\text{line } 1 * \text{line } 2)s^2 + (\text{line } 1)s + 1
\end{aligned}$$

Next, using Variation 1 only:

$$\begin{aligned}
&s^5 + \left(\frac{\cos(0)}{\sin\left(\frac{1}{10}\pi\right)}\right)s^4 \\
&\quad + \left(\frac{\cos(0)}{\sin\left(\frac{1}{10}\pi\right)}\right)\frac{\left(\frac{\cos(0)}{\sin\left(\frac{1}{10}\pi\right)}\right)}{2}s^3 \\
&\quad + \left(\frac{\cos(0)}{\sin\left(\frac{1}{10}\pi\right)}\right)\frac{\left(\frac{\cos(0)}{\sin\left(\frac{1}{10}\pi\right)}\right)}{2}s^2 \\
&\quad + \left(\frac{\cos(0)}{\sin\left(\frac{1}{10}\pi\right)}\right)s + 1
\end{aligned}$$

Table A.3 Fifth-order Butterworth polynomial factors with respect to Eq. (2.1) (variations 1–5), and derivations (variations 6–9)

Line	Factor	Variation 1	Variation 2	Variation 3	Variation 4	Variation 5	Variation 6	Variation 7	Variation 8	Variation 9
1	3.236067	$\left(\frac{\cos(0)}{\sin\left(\frac{1}{10}\pi\right)}\right)$	$2\left(\frac{\cos\left(\frac{1}{10}\pi\right)}{\sin\left(\frac{2}{10}\pi\right)}\right)$		$\frac{2}{\left(\frac{\cos\left(\frac{3}{10}\pi\right)}{\sin\left(\frac{4}{10}\pi\right)}\right)}$	$\frac{1}{\left(\frac{\cos\left(\frac{4}{10}\pi\right)}{\sin\left(\frac{5}{10}\pi\right)}\right)}$	$\left(\sqrt{5}+1\right)$	$\left(\frac{4}{(\sqrt{5}-1)}\right)$	$\csc\left(\frac{\pi}{10}\right)$	2Φ
2	1.6180339	$\frac{\left(\frac{\cos(0)}{\sin\left(\frac{1}{10}\pi\right)}\right)}{2}$	$\left(\frac{\cos\left(\frac{1}{10}\pi\right)}{\sin\left(\frac{2}{10}\pi\right)}\right)$		$\frac{1}{\left(\frac{\cos\left(\frac{3}{10}\pi\right)}{\sin\left(\frac{4}{10}\pi\right)}\right)}$	$\frac{1}{2\left(\frac{\cos\left(\frac{4}{10}\pi\right)}{\sin\left(\frac{5}{10}\pi\right)}\right)}$	$\left(\frac{\sqrt{5}+1}{2}\right)$	$\left(\frac{2}{(\sqrt{5}-1)}\right)$	$\frac{\csc\left(\frac{\pi}{10}\right)}{2}$	Φ
3	1			$\left(\frac{\cos\left(\frac{2}{10}\pi\right)}{\sin\left(\frac{3}{10}\pi\right)}\right)$						1
4	0.6180339	$\frac{2}{\left(\frac{\cos(0)}{\sin\left(\frac{1}{10}\pi\right)}\right)}$	$\frac{1}{\left(\frac{\cos\left(\frac{1}{10}\pi\right)}{\sin\left(\frac{2}{10}\pi\right)}\right)}$		$\left(\frac{\cos\left(\frac{3}{10}\pi\right)}{\sin\left(\frac{4}{10}\pi\right)}\right)$	$2\left(\frac{\cos\left(\frac{4}{10}\pi\right)}{\sin\left(\frac{5}{10}\pi\right)}\right)$	$\left(\frac{2}{\sqrt{5}+1}\right)$	$\left(\frac{\sqrt{5}-1}{2}\right)$	$\frac{2}{\csc\left(\frac{\pi}{10}\right)}$	$\frac{1}{\Phi}$
5	0.3090169	$\frac{1}{\left(\frac{\cos(0)}{\sin\left(\frac{1}{10}\pi\right)}\right)}$	$\frac{1}{2\left(\frac{\cos\left(\frac{1}{10}\pi\right)}{\sin\left(\frac{2}{10}\pi\right)}\right)}$		$\frac{\left(\frac{\cos\left(\frac{3}{10}\pi\right)}{\sin\left(\frac{4}{10}\pi\right)}\right)}{2}$	$\left(\frac{\cos\left(\frac{4}{10}\pi\right)}{\sin\left(\frac{5}{10}\pi\right)}\right)$	$\left(\frac{1}{\sqrt{5}+1}\right)$	$\left(\frac{\sqrt{5}-1}{4}\right)$	$\frac{1}{\csc\left(\frac{\pi}{10}\right)}$	$\frac{1}{2\Phi}$

which translates to:

$$s^5 + \csc\left(\frac{1}{10}\pi\right)s^4 + \frac{\csc\left(\frac{1}{10}\pi\right)^2}{2}s^3 + \frac{\csc\left(\frac{1}{10}\pi\right)^2}{2}s^2 + \csc\left(\frac{1}{10}\pi\right)s + 1$$

and to:

$$s^5 + (2\Phi)s^4 + (2\Phi^2)s^3 + (2\Phi^2)s^2 + (2\Phi)s + 1$$

In fact we can use any variation from Table A.3 to form the normalized 5th-order Butterworth Polynomial, which relieves the necessity of having to employ recursion, at least to this point.

A.1.4 6th Order

From Table A.4, the factors from Lines 1, 2, 5, and 6 may all be expressed in terms of each other. For example, the factor 3.863703 may be expressed in terms of 1.931851, 0.5176380, and 0.2588190.

In the 6th-order case, the secondary correlated factor family becomes visible as a reciprocal pair. Lines 3 and 4 evaluate to $\sqrt{\frac{3}{2}}$ and $\frac{1}{\sqrt{\frac{3}{2}}}$, respectively, and therefore remain correlated with each other through inversion. They are not, however, generated by the same primary relationship that connects the outer factor lines. Beginning at this point, the recursive product structure must therefore be retained explicitly.

Table A.4 Sixth-order Butterworth polynomial factors with respect to Eq. (2.1) (variations 1–6), and derivations (variations 7–10)

Line	Factor	Variation 1	Variation 2	Variation 3	Variation 4	Variation 5	Variation 6	Variation 7	Variation 8	Variation 9	Variation 10
1	3.863703	$\left(\frac{\cos(0)}{\sin\left(\frac{1}{12}\pi\right)}\right)$	$2\left(\frac{\cos\left(\frac{1}{12}\pi\right)}{\sin\left(\frac{2}{12}\pi\right)}\right)$			$\frac{2}{\left(\frac{\cos\left(\frac{4}{12}\pi\right)}{\sin\left(\frac{5}{12}\pi\right)}\right)}$	$\frac{1}{\left(\frac{\cos\left(\frac{5}{12}\pi\right)}{\sin\left(\frac{6}{12}\pi\right)}\right)}$	$\left(2(\sqrt{2+\sqrt{3}})\right)$	$\frac{2}{\sqrt{2-\sqrt{3}}}$		$\csc\left(\frac{\pi}{12}\right)$
2	1.931851	$\frac{\left(\frac{\cos(0)}{\sin\left(\frac{1}{12}\pi\right)}\right)}{2}$	$\left(\frac{\cos\left(\frac{1}{12}\pi\right)}{\sin\left(\frac{2}{12}\pi\right)}\right)$			$\frac{1}{\left(\frac{\cos\left(\frac{4}{12}\pi\right)}{\sin\left(\frac{5}{12}\pi\right)}\right)}$	$\frac{1}{2\left(\frac{\cos\left(\frac{5}{12}\pi\right)}{\sin\left(\frac{6}{12}\pi\right)}\right)}$	$\left((\sqrt{2+\sqrt{3}})\right)$	$\frac{1}{\sqrt{2-\sqrt{3}}}$		$\frac{\csc\left(\frac{\pi}{12}\right)}{2}$
3	1.224744			$\left(\frac{\cos\left(\frac{2}{12}\pi\right)}{\sin\left(\frac{3}{12}\pi\right)}\right)$	$\frac{1}{\left(\frac{\cos\left(\frac{3}{12}\pi\right)}{\sin\left(\frac{4}{12}\pi\right)}\right)}$					$\sqrt{\frac{3}{2}}$	
4	0.8164965			$\frac{1}{\left(\frac{\cos\left(\frac{2}{12}\pi\right)}{\sin\left(\frac{3}{12}\pi\right)}\right)}$	$\left(\frac{\cos\left(\frac{3}{12}\pi\right)}{\sin\left(\frac{4}{12}\pi\right)}\right)$					$\frac{1}{\sqrt{\frac{3}{2}}}$	
5	0.5176380	$\frac{2}{\left(\frac{\cos(0)}{\sin\left(\frac{1}{12}\pi\right)}\right)}$	$\frac{1}{\left(\frac{\cos\left(\frac{1}{12}\pi\right)}{\sin\left(\frac{2}{12}\pi\right)}\right)}$			$\left(\frac{\cos\left(\frac{4}{12}\pi\right)}{\sin\left(\frac{5}{12}\pi\right)}\right)$	$2\left(\frac{\cos\left(\frac{5}{12}\pi\right)}{\sin\left(\frac{6}{12}\pi\right)}\right)$	$\left(\frac{1}{(\sqrt{2+\sqrt{3}})}\right)$	$\sqrt{2-\sqrt{3}}$		$\frac{2}{\csc\left(\frac{\pi}{12}\right)}$
6	0.2588190	$\frac{1}{\left(\frac{\cos(0)}{\sin\left(\frac{1}{12}\pi\right)}\right)}$	$\frac{1}{2\left(\frac{\cos\left(\frac{1}{12}\pi\right)}{\sin\left(\frac{2}{12}\pi\right)}\right)}$			$\frac{\left(\frac{\cos\left(\frac{4}{12}\pi\right)}{\sin\left(\frac{5}{12}\pi\right)}\right)}{2}$	$\left(\frac{\cos\left(\frac{5}{12}\pi\right)}{\sin\left(\frac{6}{12}\pi\right)}\right)$	$\left(\frac{1}{2(\sqrt{2+\sqrt{3}})}\right)$	$\frac{\sqrt{2-\sqrt{3}}}{2}$		$\frac{1}{\csc\left(\frac{\pi}{12}\right)}$

Appendix B
MATLAB Solver Output for Design Examples

This appendix preserves the complete MATLAB candidate solution output for selected design examples in Chaps. 7, 8, 9, and 10. The MATLAB input code, solution-count discussion, selected real-valued tuples, circuit realizations, and high-pass transformations remain in the corresponding chapters; the raw candidate data are collected here so that the main chapters remain readable while the complete numerical solver output remains available.

B.1 Second-Order Low-Pass Shunt-First Solver Output

This is the complete candidate solution output corresponding to Sect. 7.2.2.

```
C1 =

0.00000000014196096919434533255739716247942
0.000000000084063139780882748385895099403302

L2 =

0.030375374928424172301759335218389
0.051296176608684742466289890690313
```


K. M. P. Ryan, *Polynomial Symmetry in Butterworth Filter Synthesis*,
Synthesis Lectures on Engineering, Science, and Technology,
https://doi.org/10.1007/978-3-032-33468-8

B.2 Second-Order Low-Pass Series-First Solver Output

This is the complete candidate solution output corresponding to Sect. 7.3.2.

```
L1 =

0.015944425907205871471791236665086
0.11671911944266447590557622403497

C2 =

0.000000000014559116870969773339655498414454
0.0000000000019888494818328394300021875843208
```

B.3 Second-Order High-Pass Shunt-First Low-Pass Precursor Solver Output

This is the complete candidate solution output corresponding to the low-pass precursor solve used in Sect. 7.5.1.

```
C1 =

0.000000000097428795207745448290705485275164
0.000000000026851928340265186060358766719906

L2 =

0.27855304346555875267876192755929
1.0106941699827327245698427714344
```

B.4 Second-Order High-Pass Series-First Low-Pass Precursor Solver Output

This is the complete candidate solution output corresponding to the low-pass precursor solve used in Sect. 7.6.1.

```
L1 =

2.7321618409437607529487541127896
3.4800207405402711269821203051569

C2 =

0.000000011697548707698390342797043042544
0.0000000091837372804832294216764844127383
```

B.5 Third-Order Low-Pass Shunt-First Solver Output

This is the complete candidate solution output corresponding to Sect. 8.2.1.

```
C1 =

0.00000000000082762918711840346950821337112456
0.00000000000090982074596637520019050146020515
0.00000000000094958117204497884491960066218075 +
0.000000000000085935532065388360199151768339271i
0.00000000000094958117204497884491960066218075 -
0.000000000000085935532065388360199151768339271i


L2 =

0.46908041897527884624147168607 63
0.47375276063214325496358018695138
0.47608893146057557023580595431692 -
0.0040463665700040139385107521491788i
0.47608893146057557023580595431692 +
0.0040463665700040139385107521491788i


C3 =

0.0000000000010081603155446093642159729791861
0.00000000000090804027483920925886614143100922
0.00000000000085931560783189100122032465347806 -
0.000000000000070409011325570187686543840212266i
0.00000000000085931560783189100122032465347806 +
0.000000000000070409011325570187686543840212266i
```

B.6 Third-Order Low-Pass Series-First Solver Output

This is the complete candidate solution output corresponding to Sect. 8.3.1.

```
L1 =

0.038259088693457829001686074659601 -
0.016851357575687219476838778265374i
0.038259088693457829001686074659601 +
0.016851357575687219476838778265374i
0.036410895444369693197557389257522
0.025912778550373908968030772722 68
```

```
C2 =

0.00000000084458289230131941286026914513528 +
0.00000000011112242389804684728711182054587i
0.00000000084458289230131941286026914513528 -
0.00000000011112242389804684728711182054587i
0.00000000078042633095077802074654423611192
0.00000000065211320824969532945087295003166


L3 =

0.024414516146562880069450927234181 +
0.0071281454846081203392164939676105i
0.024414516146562880069450927234181 -
0.0071281454846081203392164939676105i
0.031867408513823387683530542175803
0.053601923644163907401428396242668
```

B.7 Third-Order High-Pass Shunt-First Low-Pass Precursor Solver Output

This is the complete candidate solution output corresponding to the low-pass precursor solve used in Sect. 8.5.1.

```
C1 =

0.000000000037967023230675529908643691266032
0.000000000073860690993403715529503121412084
0.000000000041067276848869177731425897273632 +
0.000000000063764094326393583679582427240447i
0.000000000041067276848869177731425897273632 -
0.000000000063764094326393583679582427240447i


L2 =

0.022260941280085205051233216725535
0.20954122927377550881957006355206
0.30318137327062063216704694309569 -
0.16218948703060159164054392558373i
0.30318137327062063216704694309569 +
0.16218948703060159164054392558373i
```

```
C3 =

0.00000000011366229302409001090053507321729
0.000000000006207026958872200121816858928541
0.0000000000032196419521619362451088431727948 -
0.0000000000017899322610202109495212313987143i
0.0000000000032196419521619362451088431727948 +
0.0000000000017899322610202109495212313987143i
```

B.8 Third-Order High-Pass Series-First Low-Pass Precursor Solver Output

This is the complete candidate solution output corresponding to the low-pass precursor solve used in Sect. 8.6.1.

```
L1 =

0.04372430729808566194686710339884 4 -
0.040833484168727993215784557981341i
0.043724307298085661946867103398843 +
0.040833484168727993215784557981341i
0.053182963656945225304024579666458
0.031272950739029633745086796211709


C2 =

0.0000000000091556931256361989823125977142678 +
0.0000000000031250147551451048912630702549223i
0.0000000000091556931256361989823125977142679 -
0.0000000000031250147551451048912630702549223i
0.0000000000073514650155316204307324786910002
0.0000000000037430087953224621949490301950831


L3 =

0.015998632034968221587504669123554 +
0.0071887940611470076802235278492835i
0.015998632034968221587504669123554 -
0.0071887940611470076802235278492835i
0.025964746566522077061296649290194
0.086724301265264518476237985 11411
```

B.9 Fourth-Order Low-Pass Shunt-First Solver Output

This is the complete candidate solution output corresponding to Sect. 9.2.1.

```
C1 =

0.0000000011470144085092337192384356213433
0.00000000089233741567988272338585982494 27
0.00000000069649885088818770158043855342489 +
0.00000000022817482842556779206041513035988i
0.00000000069649885088818770158043855342489 -
0.00000000022817482842556779206041513035988i
0.00000000075730466370193903704349356423726 +
0.00000000010276430144353570066811476065201i
0.00000000075730466370193903704349356423726 -
0.00000000010276430144353570066811476065201i
0.00000000067909769912128970419295464326354
0.00000000058093402422166264813152387302468


L2 =

0.045351245077589109286992603119532
0.044364147498599375600887500405825
0.044851891184281448698552366054 99 +
0.0073370852676193773853041109782 13i
0.044851891184281448698552366054 99 -
0.0073370852676193773853041109782 13i
0.043465664586078671253199823140494 +
0.0032272276217920735024192985688374i
0.043465664586078671253199823140494 -
0.0032272276217920735024192985688374i
0.041271690102789409152346862914895
0.036716464506965000976459152352763


C3 =

0.0000000016105407809139997621003593515442
0.0000000018103524100251521718227735776966
0.0000000019673953040794403226020443404126 -
0.00000000032183586288115316460084004361043i
0.0000000019673953040794403226020443404126 +
0.00000000032183586288115316460084004361043i
0.0000000019065894912656889871389893296002 -
0.00000000014155997218093455023420441488742i
0.0000000019065894912656889871389893296002 +
0.00000000014155997218093455023420441488742i
```

```
0.0000000019460007851089314489633777417722
0.0000000019892990962903599188946469417628


L4 =

0.013243901510595776387043228647668
0.015481797705487514160309302775265
0.015878502203008547945550205965559 -
0.0052018384685147242962765199758923i
0.015878502203008547945550205965559 +
0.0052018384685147242962765199758923i
0.017264728801211325390902748880055 -
0.0023427794385891494895514130674403i
0.017264728801211325390902748880055 +
0.0023427794385891494895514130674403i
0.020343151467703694374661477945114
0.026149175679430106637710159921136


M_C1 =

0.0000000011470144085092337192384356213433
0.00000000089233741567988272338585982494 27
0.00000000073292182503702482249720271822038
0.00000000073292182503702482249720271822038
0.00000000076424528478485547995765396036 14
0.00000000076424528478485547995765396036 14
0.00000000067909769912128970419295464326354
0.00000000058093402422166264813152387302468


M_L2 =

0.045351245077589109286992603119532
0.044364147498599375600887500405825
0.045448046856063476441653189837252
0.045448046856063476441653189837252
0.043585307111827848982573334708623
0.043585307111827848982573334708623
0.041271690102789409152346862914895
0.036716464506965000976459152352763


M_C3 =

0.0000000016105407809139997621003593515442
0.0000000018103524100251521718227735776966
0.0000000019935452352907093922892405269525
```

```
0.00000000199354523529070939228924052695 25
```

```
0.0000000019935452352907093922892405269525
0.0000000019118375228895957900205870226964
0.0000000019118375228895957900205870226964
0.0000000019460007851089314489633777417722
0.0000000019892990962903599188946469417628


M_L4 =

0.013243901510595776387043228647668
0.015481797705487514160309302775265
0.016708858598464077093362228688901
0.016708858598464077093362228688901
0.017422958304411221289882611926735
0.017422958304411221289882611926735
0.020343151467703694374661477945114
0.026149175679430106637710159921136
```

B.10 Fourth-Order Low-Pass Series-First Solver Output

This is the complete candidate solution output corresponding to Sect. 9.3.1.

```
L1 =

0.0030388838675490481782638824525194
0.0041520183590461141444611548373143
0.035892132393468997416860066312083
0.0049932534125805952905297825570182 +
0.0017453820305764933747532317474346i
0.0049932534125805952905297825570182 -
0.0017453820305764933747532317474346i
0.0086145462009159563249434397980515
0.0032727168989277372561263879513452 +
0.0027617923178417275426552219565592i
0.0032727168989277372561263879513452 -
0.0027617923178417275426552219565592i


C2 =

0.00000000000450782557588332061388888292609 67
0.0000000000068453409010146469857251287560427
0.0000000000021379158931008671935611923904965
```

```
0.000000000058079752605461545375309638594922 +
0.000000000019589046917274826151751649803515i
0.000000000058079752605461545375309638594922 -
0.000000000019589046917274826151751649803515i
0.000000000042932317550455719924302412458373
0.000000000061034173433905906310819279720125 +
0.000000000044798955277023932838711760950645i
0.000000000061034173433905906310819279720125 -
0.000000000044798955277023932838711760950645i


L3 =

0.012450366995062210188422960005295
0.025002064448683393055116752919258
0.026251773023714105927043298608417
0.033823324727316585564765321132139 -
0.011407877362744167757734090779575i
0.033823324727316585564765321132139 +
0.011407877362744167757734090779575i
0.039864527271148898186068859823691
0.035543861240969443599168715737812 -
0.026089119595127657527952181107218i
0.035543861240969443599168715737812 +
0.026089119595127657527952181107218i


C4 =

0.000000000061632207557986464415241545284846
0.000000000014792475789745099809299127340565
0.0000000000052182221779467136792772210531619
0.0000000000085741696074259827092001211570476 -
0.0000000000029970843302707833208895386830047i
0.0000000000085741696074259827092001211570476 +
0.0000000000029970843302707833208895386830047i
0.0000000000071296420754277665781666921445742
0.0000000000056197487789816217736904800318449 -
0.0000000000047424141730917774961453773552304i
0.0000000000056197487789816217736904800318449 +
0.0000000000047424141730917774961453773552304i


M_L1 =

0.0030388838675490481782638824525194
0.0041520183590461141444611548373143
0.035892132393468997416860066312083
0.0052895120828775014557357859090524
```

```
0.0052895120828775014557357859090524
0.0086145462009159563249434397980515
0.0042823092727425429384891894095315
0.0042823092727425429384891894095315


M_C2 =

0.000000000045078255758833206138888829260967
0.000000000068453409010146469857251287560427
0.000000000021379158931008671935611923904965
0.000000000061294277235634417244043881884792
0.000000000061294277235634417244043881884792
0.000000000042932317550455719924302412458373
0.000000000075710743759871332876324860684187
0.000000000075710743759871332876324860684187


M_L3 =

0.012450366995062210188422960005295
0.025002064448683393055116752919258
0.026251773023714105927043298608417
0.035695335290944059226241395054428
0.035695335290944059226241395054428
0.039864527271148898186068859823691
0.044090908735998669413856545868043
0.044090908735998669413856545868043


M_C4 =

0.000000000061632207557986464415241545284846
0.000000000014792475789745099809299127340565
0.0000000000052182221779467136792772210531619
0.0000000000090828904507134787000065009771488
0.0000000000090828904507134787000065009771488
0.0000000000071296420754277665781666921445742
0.0000000000073533712355631275130317834492951
0.0000000000073533712355631275130317834492951
```

B.11 Fourth-Order High-Pass Shunt-First Solver Output

This is the complete candidate solution output for the low-pass precursor solve corresponding to Sect. 9.5.1.

```
C1 =

0.000000000035437914697207821784765410906707
0.000000000014375719413315251550107912081539
0.0000000000067017432569755488023442574329852 +
0.0000000000047875726527648964945093513420878i
0.0000000000067017432569755488023442574329852 -
0.0000000000047875726527648964945093513420878i
0.0000000000093069620935820902074886716090539 +
0.0000000000027539740979739804651006679643237i
0.0000000000093069620935820902074886716090539 -
0.0000000000027539740979739804651006679643237i
0.0000000000078106637202540160808067640408949
0.0000000000059041125790971426066604196080231


L2 =

0.0031333128132035115298523076769151
0.0041966299725687486241032862971159
0.0052721432784210758934179100487265 +
0.0028744933411979033322238894419 33i
0.0052721432784210758934179100487265 -
0.0028744933411979033322238894419 33i
0.0049289932320510740430813463333458 +
0.0012921938484213345572620482434408i
0.0049289932320510740430813463333458 -
0.0012921938484213345572620482434408i
0.0051908050599651284322901182538162
0.0037310097331681333977549054576196


C3 =

0.00000000002832608341230198667557650723539
0.000000000039408950289904130634379001049347
0.000000000040026475658468314449969100649019 -
0.000000000021823351846830792006340056681879i
0.000000000040026475658468314449969100649019 +
0.000000000021823351846830792006340056681879i
0.000000000037421256821861773044824686472951 -
0.0000000000098104248857494932845268185544152i
0.000000000037421256821861773044824686472951 +
```

```
0.00000000000098104248857494932845268185544152i
0.0000000000031861104407414328239333038992228
0.0000000000023788327142280775437624378413889


L4 =

0.00077766845411339087443592649325821
0.0010287925068424660837059760551161
0.00088272949553309417626909714974606 -
0.00063060183456064221262939152511497i
0.00088272949553309417626909714974606 +
0.00063060183456064221262939152511497i
0.0012258795419030960266056608651268 -
0.00036274355387838000053338562185605i
0.0012258795419030960266056608651268 +
0.00036274355387838000053338562185605i
0.0018935180085319969992746337908969
0.0046677545474233043373308747691522
```

B.12 Fourth-Order High-Pass Series-First Solver Output

This is the complete candidate solution output for the low-pass precursor solve corresponding to Sect. 9.6.1.

```
L1 =

0.039674374976236741293023013586585
0.032921182426271299358578886917255
0.029049190641253742089029132767754 +
0.0031207593687437667487875929851544i
0.029049190641253742089029132767754 -
0.0031207593687437667487875929851544i
0.026533911792973501578715817548747
0.027532423138075666107666569428854 +
0.0071407917388190413973122829362425i
0.027532423138075666107666569428854 -
0.0071407917388190413973122829362425i
0.023332858905168789011627503218721
```

```
C2 =

0.000000061989928745454762865072364882869
0.000000060211807049229028728012335507487
0.000000058748266758089938065996822546195 +
0.0000000031523796245230867287229435984005i
0.000000058748266758089938065996822546195 -
0.0000000031523796245230867287229435984005i
0.000000056173577460613978641297879521488
0.000000059990500585426609631404306456992 +
0.0000000072883326685752283470698010327863i
0.000000059990500585426609631404306456992 -
0.0000000072883326685752283470698010327863i
0.000000051486250471873314637443617388185


L3 =

0.062864711826157317172318656830734
0.068587938079409667921024710895672
0.07173163371162781437858212032887 -
0.0038490555215426888957707141336267i
0.07173163371162781437858212032887 +
0.0038490555215426888957707141336267i
0.073518616407108644076903061654949
0.073248401214805890359944658184052 -
0.008899054188330353811772227061097i
0.073248401214805890359944658184052 +
0.008899054188330353811772227061097i
0.075689702998200265458253357522023


C4 =

0.000000019109630552963791164314099964228
0.000000021731295489740787533755865000972
0.000000023791310926497741268656159789191 -
0.000000002555904478905623872880914545172i
0.000000023791310926497741268656159789191 +
0.000000002555904478905623872880914545172i
0.000000026962475369591563766239907443107
0.000000022549077099161069703248693011311 -
0.0000000058483142824070773114760586007879i
0.000000022549077099161069703248693011311 +
0.0000000058483142824070773114760586007879i
0.000000032493345598883490002475896427872
```

B.13 Fifth-Order Low-Pass Shunt-First Solver Output

This is the complete candidate solution output corresponding to Sect. 10.2.1.

```
C1 =

0.00000000000068179853099117962461972237286078 +
0.0000000000001256810304423797153240805689126i
0.00000000000068179853099117962461972237286078 -
0.0000000000001256810304423797153240805689126i
0.00000000000063582387152124274846385238345822 +
0.000000000000066037571068458325506061317204136i
0.00000000000063582387152124274846385238345822 -
0.000000000000066037571068458325506061317204136i
0.00000000000062170227612494750088303367575108
0.0000000000005743717312886265971392449550086 +
0.000000000000053781754705220462462844183203416i
0.0000000000005743717312886265971392449550086 -
0.000000000000053781754705220462462844183203416i
0.00000000000056230014467760369742519007749969
0.00000000000052629787425008178242565215617378 +
0.000000000000073854613187009889302931520662467i
0.00000000000052629787425008178242565215617378 -
0.000000000000073854613187009889302931520662467i
0.00000000000052053868696599583000602273149802 +
0.000000000000027136558710073201566831220669907i
0.00000000000052053868696599583000602273149802 -
0.000000000000027136558710073201566831220669907i
0.00000000000049768268439179066449774110573237 +
0.0000000000001099049395144945264615739693552 7i
0.00000000000049768268439179066449774110573237 -
0.0000000000001099049395144945264615739693552 7i
0.00000000000048700895517591736332070878274079
0.0000000000004497873134375272270417789008073


L2 =

0.43357123454482784553766001937093 +
0.030420216775111290148494914166917i
0.43357123454482784553766001937093 -
0.030420216775111290148494914166917i
0.41421602712348684840021806856865 +
0.018800727911486029568954293464281i
0.41421602712348684840021806856865 -
0.018800727911486029568954293464281i
```

```
0.40796329663456960153130677336133
0.38874972771165212005855565975927 +
0.01995512267101548601569119985156i
0.38874972771165212005855565975927 -
0.01995512267101548601569119985156i
0.38289864365646085845284172806061
0.36967370643140186047187007449728 +
0.035293369251165555265857780191378i
0.36967370643140186047187007449728 -
0.035293369251165555265857780191378i
0.36476910039047347982439754096217 +
0.014625132010995761140968373026723i
0.36476910039047347982439754096217 -
0.014625132010995761140968373026723i
0.36150024572855309409592690011286 +
0.057105871025213096582083282644054i
0.36150024572855309409592690011286 -
0.057105871025213096582083282644054i
0.34787377262241326401360442976631
0.32622721187098027624904383870104


C3 =

0.0000000000017903784095492359756435707781793 -
0.000000000000062676731705047568433185247796009i
0.0000000000017903784095492359756435707781793 +
0.000000000000062676731705047568433185247796009i
0.0000000000018041999863579982473700051853505 -
0.000000000000021826459054423137959928582016435i
0.0000000000018041999863579982473700051853505 +
0.000000000000021826459054423137959928582016435i
0.0000000000018030046202997697764173177070343
0.0000000000018247386751678534353118848333244 -
0.0000000000000018563000886635535070981050607682i
0.0000000000018247386751678534353118848333244 +
0.0000000000000018563000886635535070981050607682i
0.0000000000018178610208055852363779904413656
0.0000000000018416313741554445558818715178 18 +
0.000000000000021797521760821708560317298406586i
0.0000000000018416313741554445558818715178 18 -
0.000000000000021797521760821708560317298406586i
0.0000000000018184115446109908330494852668058 +
0.000000000000012332277982177015132223142651688i
0.0000000000018184115446109908330494852668058 -
0.000000000000012332277982177015132223142651688i
0.0000000000018769353789197706534075610951776 +
0.000000000000048091350052717530094918496040103i
0.0000000000018769353789197706534075610951776 -
```

```
0.00000000000004809135005271753009491849604010 3i
0.000000000001797642825094034373460678016 02
0.000000000001754677527089911376672939131015 6


L4 =

0.33540498963794533647505378856734 -
0.03218437480399254747388631214242 4i
0.33540498963794533647505378856734 +
0.03218437480399254747388631214242 4i
0.35383272429678166119839993641796 -
0.01989103753485382899453198998803 6i
0.35383272429678166119839993641796 +
0.01989103753485382899453198998803 6i
0.35973119171405551461611050670082
0.37815260432689183607930025494836 -
0.02104543229438328544126881234079 3i
0.37815260432689183607930025494836 +
0.02104543229438328544126881234079 3i
0.38364942531043970423381325475435
0.39630115284463742325189008682672 -
0.03705752728004681259124916185563 9i
0.39630115284463742325189008682672 +
0.03705752728004681259124916185563 9i
0.40085149581392241044816185623902 -
0.01529898041650921904078216812327 8i
0.40085149581392241044816185623902 +
0.01529898041650921904078216812327 8i
0.40412035047584279617663250580467 -
0.05996033867746215333305233999094i
0.40412035047584279617663250580467 +
0.05996033867746215333305233999094i
0.41681935081947795384485884213239
0.43731949218918638814865741126989


C5 =

0.000000000000459866779006144167696810238036 26 -
0.000000000000056234957465388409791814868387407i
0.000000000000459866779006144167696810238036 26 +
0.000000000000056234957465388409791814868387407i
0.00000000000049557871498263542311467865261285 -
0.000000000000040027429026552744717469758276514i
0.00000000000049557871498263542311467865261285 +
0.000000000000040027429026552744717469758276514i
0.00000000000051225503744265928919125925761193
0.00000000000054089115102503705055773604022655 -
```

```
0.00000000000004774177162907446612708420619615 1i
0.0000000000005408911510250370505577360402265 5 +
0.00000000000004774177162907446612708420619615 1i
0.0000000000005611997530038282967487614448917 2
0.0000000000005756311623913073956897735528098 5 -
0.00000000000008888279367588786076416747296416 2i
0.0000000000005756311623913073956897735528098 5 +
0.00000000000008888279367588786076416747296416 2i
0.0000000000006059695402253472184848641882009 7 -
0.00000000000003688317840778892242863498320274 9i
0.0000000000006059695402253472184848641882009 7 +
0.00000000000003688317840778892242863498320274 9i
0.0000000000005703017084907725636350677286456 1 -
0.0000000000001470432653077858766287473195332 5i
0.0000000000005703017084907725636350677286456 1 +
0.0000000000001470432653077858766287473195332 5i
0.0000000000006638268448476987957474266131188 3
0.0000000000007484127692098525928744233650124 7
```

B.14 Fifth-Order Low-Pass Series-First Solver Output

This is the complete candidate solution output corresponding to Sect. 10.3.1.

```
L1 =

1.1977235173759104381757035777101 +
0.15838733993714251949786018133392i
1.1977235173759104381757035777101 -
0.15838733993714251949786018133392i
1.1328356962377839894563534727285 +
0.086567147150021927813335921909092i
1.1328356962377839894563534727285 -
0.086567147150021927813335921909092i
1.1118446589117735163875120856666
1.0494557658408118347783183780661 +
0.074231167289232872790002146755518i
1.0494557658408118347783183780661 -
0.074231167289232872790002146755518i
1.0310110970903883090017593946 03
0.98338705721895785539118727660158 +
0.10615956760855935457495427718124i
0.98338705721895785539118727660158 -
0.10615956760855935457495427718124i
```

```
0.94608513732085099392795649712151 +
0.16174780278151080984871233575282i
0.94608513732085099392795649712151 -
0.16174780278151080984871233575282i
0.97215488272593935612313716010321 +
0.039235637312100377931889063108986i
0.97215488272593935612313716010321 -
0.039235637312100377931889063108986i
0.92249386043226118526469628090213
0.86615069962976453599037288322496


C2 =

0.00000073933654818780484047285206160512 +
0.000000043327203210109964851295986440127i
0.00000073933654818780484047285206160512 -
0.000000043327203210109964851295986440127i
0.00000071417315685841114255439602652972 +
0.000000026777684208814822729341934804036i
0.00000071417315685841114255439602652972 -
0.000000026777684208814822729341934804036i
0.00000070546589994009797719232436583896
0.00000068186311324265071800780095871968 +
0.000000027669424436505887684347727135779i
0.00000068186311324265071800780095871968 -
0.000000027669424436505887684347727135779i
0.00000067345793441046954166688856035649
0.00000065682386222829420711405625700201 +
0.000000047100349753834960965212026649134i
0.00000065682386222829420711405625700201 -
0.000000047100349753834960965212026649134i
0.00000064654344026944974475475540200774 +
0.000000076209966771203695153789452574084i
0.00000064654344026944974475475540200774 -
0.000000076209966771203695153789452574084i
0.00000064916257652876023311312903188938 +
0.000000018879850229471409915216904360261i
0.00000064916257652876023311312903188938 -
0.000000018879850229471409915216904360261i
0.00000062581564872637877976709198851623
0.00000059637109685409063034341045997689


L3 =

3.3089681554735813146416252973585 -
0.068455949445375030630311466404317i
3.3089681554735813146416252973585 +
```

```
0.06845594944537503063031l466404317i
3.32350664735651576537l4083997971 -
0.02482264041056251370246414983971 9i
3.32350664735651576537140839979 71 +
0.0248226404105625137024641498397 19i
3.3218911788133480638688683711867
3.3435386337617486815133247492091 -
0.0038124507952593944973170136186667i
3.3435386337617486815133247492091 +
0.0038124507952593944973170136186667i
3.3355887234936126674471978081967
3.3599147909999311974799008255666 +
0.020503928425143173689992552110 12i
3.3599147909999311974799008255666 -
0.02050392842514317368999255211012i
3.3973626129363829416280182634486 +
0.048062075050680534754020073480844i
3.3973626129363829416280182634486 -
0.048062075050680534754020073480844i
3.3344875829293083744120708 03892 +
0.011964856643915352647198260451049i
3.3344875829293083744120708 03892 -
0.011964856643915352647198260451049i
3.3111007935274202462286420352642
3.2652820806481337928784828969376


C4 =

0.0000006093464151981342535606221 7324335 -
0.00000004436654691167018304538595 3201734i
0.0000006093464151981342535606221 7324335 +
0.00000004436654691167018304538595 3201734i
0.0000006339633912072156915322060307589 -
0.000000027420033942341697045094321540884i
0.0000006339633912072156915322060307589 +
0.000000027420033942341697045094321540884i
0.00000064246193604511589680055262225408
0.0000006655980283431724194496439907 0161 -
0.000000028311774170032762000100267947023i
0.0000006655980283431724194496439907 0161 +
0.000000028311774170032762000100267947023i
0.0000006737944950949406356968312 0893564
0.0000006900908640372166703965166103 8555 -
0.000000048139693455395179159302224522334i
0.0000006900908640372166703965166103 8555 +
0.000000048139693455395179159302224522334i
0.0000007001625739156481726620922 1129495 -
0.000000077891660206290787663631858457834i
```

```
0.00000070016257391564817266209221129495 +
0.00000007789166020629078766363185845 7834i
0.00000069754343765633768430371856908736 -
0.00000001927684419750475379355441351 0797i
0.00000069754343765633768430371856908736 +
0.00000001927684419750475379355441351 0797i
0.00000072034395013840687770288378927177
0.00000074911309553089133049740896182683


L5 =

0.88467353266758403642680955167157 -
0.08576777754462985166158815660412 7i
0.88467353266758403642680955167157 +
0.08576777754462985166158815660412 7i
0.93721180278877758838806642065905 -
0.05917125242225128854821896468410 9i
0.93721180278877758838806642065905 +
0.05917125242225128854821896468410 9i
0.96065440966951424192083075270346
1.0032654264894233952279820973224 -
0.06784546217676535273003251333398 4i
1.0032654264894233952279820973224 +
0.06784546217676535273003251333398 4i
1.0304961065195414140320504087465
1.0551469187390964126202262395025 -
0.12249988308656489105898639614813i
1.0551469187390964126202262395025 +
0.12249988308656489105898639614813i
1.0558371177123100088967233400185 -
0.20307301056784558183411978005865i
1.0558371177123100088967233400185 +
0.20307301056784558183411978005865i
1.0926424023142962139174899850655 -
0.04961013532608621893577966026790 7i
1.0926424023142962139174899850655 +
0.04961013532608621893577966026790 7i
1.1678791548758640669310501550474
1.2727467082665537378593274681123


M_L1 =

1.2081507246729138038402585349531
1.2081507246729138038402585349531
1.1361384535505513590425548161376
1.1361384535505513590425548161376
1.1118446589117735163875120856666
```

```
1.0520777873587327842720279931166
1.0520777873587327842720279931166
1.031011097090388309001759394603
0.98910058037623159389312688897711
0.98910058037623159389312688897711
0.95981218931833742276570417599528
0.95981218931833742276570417599528
0.97294632495486182707799565753419
0.97294632495486182707799565753419
0.92249386043226118526469628090213
0.86615069962976453599037288322496


M_C2 =

0.00000074060500810099065396399934765601
0.00000074060500810099065396399934765601
0.00000071467499071164910681737398157483
0.00000071467499071164910681737398157483
0.00000070546589994009797719232436583896
0.00000068242428316232081903927980789944
0.00000068242428316232081903927980789944
0.00000067345793441046954166688856035649
0.00000065851046228547256901165310777108
0.00000065851046228547256901165310777108
0.00000065101949217417714827210808861316
0.00000065101949217417714827210808861316
0.00000064943706354822848065327028029343
0.00000064943706354822848065327028029343
0.00000062581564872637877976709198851623
0.00000059637109685409063034341045997689


M_L3 =

3.3096761882324232513465405491989
3.3096761882324232513465405491989
3.3235993438589886284696504639938
3.3235993438589886284696504639938
3.3218911788133480638688683711867
3.3435408073236443234799562458594
3.3435408073236443234799562458594
3.3355887234936126674471978081967
3.3599773531768000197865726652056
3.3599773531768000197865726652056
3.3977025600891118300242608008274
3.3977025600891118300242608008274
3.3345090490961710147486139506041
3.3345090490961710147486139506041
```

```
3.3111007935274202462286420352642
3.2652820806481337928784828969376


M_C4 =

0.00000061095944562604353787838117356578
0.00000061095944562604353787838117356578
0.00000063455609653706139200603542664114
0.00000063455609653706139200603542664114
0.00000064246193604511589680055262225408
0.00000066619988884040913633448401130835
0.00000066619988884040913633448401130835
0.00000067379449509494063569683120893564
0.00000069176790234414004784615305271381
0.00000069176790234414004784615305271381
0.00000070448189518388459060575788191651
0.00000070448189518388459060575788191651
0.00000069780974781070229071513399040471
0.00000069780974781070229071513399040471
0.00000072034395013840687770288378927177
0.00000074911309553089133049740896182683


M_L5 =

0.88882133810315782263904282006638
0.88882133810315782263904282006638
0.9390778457613662145365587289573
0.9390778457613662145365587289573
0.96065440966951424192083075270346
1.005556822226861288440433300444
1.005556822226861288440433300444
1.0304961065195414140320504087465
1.062234080361212754178780224756
1.062234080361212754178780224756
1.0751886656583237555989661960464
1.0751886656583237555989661960464
1.0937680672163678284973237565693
1.0937680672163678284973237565693
1.1678791548758640669310501550474
1.2727467082665537378593274681123
```

B.15 Fifth-Order High-Pass Shunt-First Solver Output

This is the complete candidate solution output corresponding to Sect. 10.5.1.

```
C1 =

0.00000000016730737921958609666732678362305 -
0.00000000000489048408753465488635604957043 81i
0.00000000016730737921958609666732678362305 +
0.0000000000048904840875346548863560495704381i
0.00000000013229551687521119482862272801656 -
0.0000000000295794889111655079918114096490 19i
0.00000000013229551687521119482862272801656 +
0.000000000029579488911165507991811409649019i
0.00000000020066040017150341350115370642083
0.00000000044540478565645825853711267146322
0.00000000011726919091202968320151254389076 -
0.00000000013672385306792347487226364773552i
0.00000000011726919091202968320151254389076 +
0.00000000013672385306792347487226364773552i
0.00000000025719231812135850084309168636156 -
0.000000000070787313791364746953926829282105i
0.00000000025719231812135850084309168636156 +
0.000000000070787313791364746953926829282105i
0.0000000001068402001351447203459688166811 -
0.000000000032475070885000282227010386609616i
0.0000000001068402001351447203459688166811 +
0.000000000032475070885000282227010386609616i
0.00000000014979105928777138183868172211968
0.0000000001809210719835066865614217537468 5
0.00000000017310825936922628836940908062738 +
0.00000000010896819663160082348034104045984i
0.00000000017310825936922628836940908062738 -
0.00000000010896819663160082348034104045984i


L2 =

0.027851160639927614219234468546633 -
0.0070398693580900261379209839398659i
0.027851160639927614219234468546633 +
0.0070398693580900261379209839398659i
0.019310510033961169619528193990648 -
0.0028346276040930348762176348695204i
```

```
0.019310510033961169619528193990648 +
0.0028346276040930348762176348695204i
0.025741451644412445381932465400222
0.014846480550904362673595418225053
0.021977711549435019287639645087759 -
0.022706934509440564989678271358343i
0.021977711549435019287639645087759 +
0.022706934509440564989678271358343i
0.020852245150893283562884386688954 -
0.0060833736349325984255823836023029i
0.020852245150893283562884386688954 +
0.0060833736349325984255823836023029i
0.015335969539036277637745502530334 -
0.0045865238088712210959045981769294i
0.015335969539036277637745502530334 +
0.0045865238088712210959045981769294i
0.018782017030359208115040077854784
0.0081130181086687925549496447996214
0.023415332340895640311002355039979 +
0.014033657307152203042709277896935i
0.023415332340895640311002355039979 -
0.014033657307152203042709277896935i


C3 =

0.0000000013280919282136903333006879829663 +
0.00000000033499814830115156848312985153147i
0.0000000013280919282136903333006879829663 -
0.00000000033499814830115156848312985153147i
0.00000000054822048720379609870478007456111 +
0.000000000016758142507142741299365676893826i
0.00000000054822048720379609870478007456111 -
0.000000000016758142507142741299365676893826i
0.0000000014610390645314152109921665551588
0.0000000023782954376288494652482821643703
0.0000000023206152993250915214225810569496 +
0.0000000012878108515907967216607780262518i
0.0000000023206152993250915214225810569496 -
0.0000000012878108515907967216607780262518i
0.0000000020255331807464680573165322610851 +
0.00000000031822563626271989773877998693434i
0.0000000020255331807464680573165322610851 -
0.00000000031822563626271989773877998693434i
0.00000000050188182688245276558372288528296 -
0.0000000001422317571457783071300327605844i
0.00000000050188182688245276558372288528296 +
0.0000000001422317571457783071300327605844i
0.00000000058652293930749993493247618751818
```

```
0.0000000016602010702419953258201652339555
0.0000000019725331302774811675408573501057 -
0.0000000007536438296024415924805477064954 2i
0.0000000019725331302774811675408573501057 +
0.0000000007536438296024415924805477064954 2i


L4 =

0.0076859847523717272186051923665534 -
0.0004727354681345604825193754203916i
0.0076859847523717272186051923665534 +
0.0004727354681345604825193754203916i
0.0241258555387434148054455090861111 -
0.0046779772221315517442227245049239i
0.0241258555387434148054455090861111 +
0.0046779772221315517442227245049239i
0.0073547004698321299577830433600967
0.0054684648268253884278057252034032
0.0047278371965521439329186810066283 +
0.0030386797303380712571228608899653i
0.0047278371965521439329186810066283 -
0.0030386797303380712571228608899653i
0.0058533035950938796576739394054336 +
0.0014403285082792779339076918544405i
0.0058533035950938796576739394054336 -
0.0014403285082792779339076918544405i
0.0344909994019257189063853835991 39 -
0.0075691261440066860162104554297 4i
0.0344909994019257189063853835991 39 +
0.0075691261440066860162104554297 4i
0.0222133552642906102118094731094
0.0043027070886557155593174564322977
0.0057312096831462890076801232072761 -
0.0018780073542742959305942267888147i
0.0057312096831462890076801232072761 +
0.0018780073542742959305942267888147i


C5 =

0.0000000006942548951517081924695100328862 9 +
0.0000000002505213230188083104118848841912 6i
0.0000000006942548951517081924695100328862 9 -
0.0000000002505213230188083104118848841912 6i
0.0000000009449081856198885441088332701390 2 +
0.0000000005494359768486360967238999710365 4i
0.0000000009449081856198885441088332701390 2 -
0.0000000005494359768486360967238999710365 4i
```

```
0.00000000070231140060026357638164397738652
0.00000000045325398034036192268345264112267
0.00000000038259804422730471904781225712075 +
0.00000000025378834284159059125115779779639i
0.00000000038259804422730471904781225712075 -
0.00000000025378834284159059125115779779639i
0.00000000053775703559659936551228191051449 +
0.000000000084207758003882027191910538623681i
0.00000000053775703559659936551228191051449 -
0.000000000084207758003882027191910538623681i
0.00000000056023049352005462891560267133052 +
0.0000000010429675389506290974083547583666i
0.00000000056023049352005462891560267133052 -
0.0000000010429675389506290974083547583666i
0.0000000010634668538218220993085176000481
0.00000000037177242643469623982975343385623
0.0000000005004844820995208893241999877374 -
0.00000000022358507794900973900801127025873i
0.0000000005004844820995208893241999877374 +
0.00000000022358507794900973900801127025873i


M_C1 =

0.00000000017430846968053318390764935813843
0.00000000017430846968053318390764935813843
0.00000000013556197825911605245996713027231
0.00000000013556197825911605245996713027231
0.00000000020066040017150341350115370642083
0.00000000044540478565645825853711267146322
0.00000000018012627552609084198572715322185
0.00000000018012627552609084198572715322185
0.00000000026675594144167662756417118635527
0.00000000026675594144167662756417118635527
0.00000000011166673002243582728662220055584
0.00000000011166673002243582728662220055584
0.00000000014979105928777138183868172211968
0.0000000018092107198350668656142175374685
0.00000000020454959628165131654128348176839
0.00000000020454959628165131654128348176839


M_L2 =

0.028727111055064831044359289773769
0.028727111055064831044359289773769
0.019517451458261686465135986457 61
0.019517451458261686465135986457 61
0.025741451644412445381932465400222
```

```
0.014846480550904362673595418225053
0.031601023397450157073948407939261
0.031601023397450157073948407939261
0.021721500008405163396107033016252
0.021721500008405163396107033016252
0.01600712848550892189274888953705
0.01600712848550892189274888953705
0.018782017030359208115040077854784
0.0081130181086687925549496447996214
0.027298742206357781743745972550146
0.027298742206357781743745972550146


M_C3 =

0.0000000013696904501205950720604960804838
0.0000000013696904501205950720604960804838
0.00000000054847656096706381168842608923596
0.00000000054847656096706381168842608923596
0.0000000014610390645314152109921665551588
0.0000000023782954376288494652482821643703
0.0000000026539992759864681576897595222136
0.0000000026539992759864681576897595222136
0.000000002050378555750063668113411613444
0.000000002050378555750063668113411613444
0.00000000052164666288172873349425037435133
0.00000000052164666288172873349425037435133
0.00000000058652293930749993493247618751818
0.0000000016602010702419953258201652339555
0.0000000021116027022004192610335091475349
0.0000000021116027022004192610335091475349


M_L4 =

0.0077005091024245327504281432887453
0.0077005091024245327504281432887453
0.024575198399343553117684963541687
0.024575198399343553117684963541687
0.0073547004698321299577830433600967
0.0054684648268253884278057252034032
0.0056201440426976148024997130055382
0.0056201440426976148024997130055382
0.0060279108477233591468219350437363
0.0060279108477233591468219350437363
0.035311764474853502038916399293422
0.035311764474853502038916399293422
0.0222133552642906102118094731094
0.0043027070886557155593174564322977
```

```
0.0060310592813284572546643804533174
0.0060310592813284572546643804533174


M_C5 =

0.00000000073807234925121224873507667720309
0.00000000073807234925121224873507667720309
0.0000000010930376809182219242607933110421
0.0000000010930376809182219242607933110421
0.00000000070231140060026357638164397738652
0.00000000045325398034036192268345264112267
0.00000000045911848841975351684806935279749
0.00000000045911848841975351684806935279749
0.00000000054431018348151690145604987113561
0.00000000054431018348151690145604987113561
0.0000000011839085662222636333909672636829
0.0000000011839085662222636333909672636829
0.0000000010634668538218220993085176000481
0.00000000037177242643469623982975343385623
0.00000000054815600325444800015142353204561
0.00000000054815600325444800015142353204561
```

B.16 Fifth-Order High-Pass Series-First Solver Output

This is the complete candidate solution output corresponding to Sect. 10.6.1.

```
L1 =

0.020256000833317175354826916681129 -
0.0045773681797106083175436388915392i
0.020256000833317175354826916681129 +
0.0045773681797106083175436388915392i
0.025670393247575738219244569062172 -
0.0076087742137675095117556570506224i
0.025670393247575738219244569062172 +
0.0076087742137675095117556570506224i
0.026459397362564730337416683324061 -
0.016967844511106271915617331688248i
0.026459397362564730337416683324061 +
0.016967844511106271915617331688248i
0.017722142844994574944003190638904 -
```

```
0.0211067224076078186916355938778811i
0.017722142844994574944003190638904 +
0.0211067224076078186916355938778811i
0.039720093676444104475420354464182 +
0.0111673958023064046042706336619351i
0.039720093676444104475420354464182 -
0.0111673958023064046042706336619351i
0.318182160695912819907327386170970
0.069915075476408328023704470005455
0.016322450583717596956937871662515 +
0.005005832799464671078978560935511 2i
0.016322450583717596956937871662515 -
0.005005832799464671078978560935511 2i
0.022977474379483844588218051767306
0.030901707804266957721934972203968


C2 =

0.0000000604522186708827825585009529 07172 -
0.0000000089367020678528721574204768039484i
0.000000060452218670882782558500952907172 +
0.0000000089367020678528721574204768039484i
0.000000087897905447097787215580974833436 -
0.0000000229021350662280393575401549232 32i
0.000000087897905447097787215580974833436 +
0.000000022902135066228039357540154923232i
0.000000072623334421504069202027314856042 -
0.000000044690335255258878573901693372104i
0.000000072623334421504069202027314856042 +
0.000000044690335255258878573901693372104i
0.000000067507009932621991506853268800104 -
0.000000072310481411636511820117531774561i
0.000000067507009932621991506853268800104 +
0.000000072310481411636511820117531774561i
0.000000064358990205197070479040887844 5 +
0.000000019123704891015440957184490573116i
0.000000064358990205197070479040887844 5 -
0.000000019123704891015440957184490573116i
0.000000022176397394256247269611038445525
0.000000044607475141941886449997672705198
0.000000047813730967107429023234583904572 +
0.000000014459887693117446965574100876442i
0.000000047813730967107429023234583904572 -
0.000000014459887693117446965574100876442i
0.000000058638032337262228221179215445 7694
0.000000080901721341988895486235364806181
```

```
L3 =

0.083857673716334512728406168478791 +
0.0029823003588336551956646220700927i
0.083857673716334512728406168478791 -
0.0029823003588336551956646220700927i
0.22518970964369405771318945118533 +
0.060502093720138409223203246386822i
0.22518970964369405771318945118533 -
0.060502093720138409223203246386822i
0.34188807627972360068310766225455 +
0.13536509930158289611855807537999i
0.34188807627972360068310766225455 -
0.13536509930158289611855807537999i
0.40323777124834216707899303509517 +
0.23203784885478695694079638428686i
0.40323777124834216707899303509517 -
0.23203784885478695694079638428686i
0.35358117089275672117838386214252 -
0.056577866804313534849329839179704i
0.35358117089275672117838386214252 +
0.056577866804313534849329839179704i
0.29018076777215809070335449945357
0.42139221586198640994796310856143
0.076672014087101939041354299913522 +
0.022137953518621107048156800785369i
0.076672014087101939041354299913522 -
0.022137953518621107048156800785369i
0.090178887709372509693570393808658
0.25000006768860892565047294689207


C4 =

0.0000000742764418825677368901960660073549 -
0.0000000154762231356849208044657496503891
0.0000000742764418825677368901960660073549 +
0.0000000154762231356849208044657496503891
0.0000000211614864716784485650180176635571 -
0.0000000015107901373097536043460715310968i
0.0000000211614864716784485650180176635571 +
0.0000000015107901373097536043460715310968i
0.0000000156691829386447518956017938430 9 +
0.0000000051893925111257183346323282331479i
0.0000000156691829386447518956017938430 9 -
0.0000000051893925111257183346323282331479i
0.000000012853267186236711653836043423207 +
0.0000000083966134639655586189619401812793i
0.000000012853267186236711653836043423207 -
```

```
0.0000000083966134639655586189619401812793i
0.000000016001286913661632681648424378811 -
0.0000000040356873504200736798013518884876i
0.000000016001286913661632681648424378811 +
0.0000000040356873504200736798013518884876i
0.000000011747736531300757540010363496085
0.000000014985927418289402027721755720237
0.0000000107681804144970505108432318874 +
0.0000000025041055051015713273695264262344i
0.0000000107681804144970505108432318874 -
0.0000000025041055051015713273695264262344i
0.0000000068158387974898119299965068047337
0.0000000020225430335497222357423831215007


L5 =

0.16598889665643455722332811076615 +
0.099246768635028124969423922639068i
0.16598889665643455722332811076615 -
0.099246768635028124969423922639068i
0.12191954285351358404651928161339 +
0.044758381307780272136097316481118i
0.12191954285351358404651928161339 -
0.044758381307780272136097316481118i
0.087499670337004707690308491473792 +
0.039606516186056016754136716864077i
0.087499670337004707690308491473792 -
0.039606516186056016754136716864077i
0.066616190851116768435595797221613 +
0.04472434534350467455546157596415i
0.066616190851116768435595797221613 -
0.04472434534350467455546157596415i
0.094274840375252684804787806348983 -
0.014941599160374338864473349220751i
0.094274840375252684804787806348983 +
0.014941599160374338864473349220751i
0.06495774924958939325218177845795
0.079336311840568431218803979580303
0.094040608300757050576389489157359 -
0.18514755729461841908421282227727i
0.094040608300757050576389489157359 +
0.18514755729461841908421282227727i
0.18867517008239036277253193625358
0.12360683121706796835430456866815
```

Index

K. M. P. Ryan, *Polynomial Symmetry in Butterworth Filter Synthesis*,
Synthesis Lectures on Engineering, Science, and Technology,
https://doi.org/10.1007/978-3-032-33468-8

Kolibri 360 GmbH
Ferdinand-Jühlke-Straße 7
99095 Erfurt, Deutschland
produktsicherheit@kolibri360.de